INCORPORATING AND COMMUNICATING CLIMATE CHANGE UNCERTAINTIES IN ENVIRONMENTAL ASSESSMENTS

INCORPORATING AND COMMUNICATING CLIMATE CHANGE UNCERTAINTIES IN ENVIRONMENTAL ASSESSMENTS

Julian Scott Yeomans

iUniverse, Inc.
New York Lincoln Shanghai

INCORPORATING AND COMMUNICATING CLIMATE CHANGE UNCERTAINTIES IN ENVIRONMENTAL ASSESSMENTS

iUniverse, Inc.

For information address:
iUniverse, Inc.
2021 Pine Lake Road, Suite 100
Lincoln, NE 68512
www.iuniverse.com

ISBN: 0-595-33295-1 (pbk)
ISBN: 0-595-66826-7 (cloth)

Printed in the United States of America

Contents

List of Illustrations

List of Tables

Preface

A changing climate produces significant implications for projects that affect the environment and consequently must be addressed in their Environmental Assessments (EAs). However, recent studies indicate that climate change impacts have been inadequately considered in project EAs and the corresponding uncertainties have been addressed even more poorly. Recognizing these deficiencies, this book investigates how the various uncertainties from climate change can be identified, analyzed, incorporated, and communicated within project EAs. The three fundamental analytical approaches considered are: scenario analysis, sensitivity analysis, and probabilistic analysis. Various concepts arising from the use of these methods will be illustrated and articulated via a significant set of examples based upon a proposed hydroelectric project. Furthermore, once the uncertainties have been addressed, it is essential for them to be clearly communicated to the disparate stakeholders of an EA. Several recommendations for effectively incorporating, analyzing, and communicating the uncertainties of climate change within EAs are provided.

Acknowledgments

To Sara, Sebastian and Mr. C. This book is for all of you.

1

Overview of Climate Change Uncertainties in Environmental Assessments

1.1 Introduction

Climate change, sometimes referred to as global warming, has become an increasingly more important concern at both the regional and global policy levels. Since ratification by Canada of the Kyoto Protocol in 2002, the debate about climate change has shifted from whether or not Canada should be addressing it to exactly how the country should proceed in directly addressing the issue (Canada 2002). Climate change can have significant implications for the environment and for projects that affect the environment. Recognizing these implications, this study provides an investigation into methods for addressing and communicating the inherent uncertainties that surround the various climate change issues that can arise within environmental assessments (EAs). Hence, the primary focus of this research project is on *uncertainties* about climate change.

1.2 Climate Change in Environmental Assessments

Prior to addressing the uncertainties that arise from climate change, it is first necessary to address climate change itself. Climate change, which is often considered by laypersons to be synonymous with global warming, has become one of most visible and studied global environmental issues of the past two decades. While *weather* refers to the day-to-day changes in the state of the atmosphere at a specific location, *climate* is the average regional weather as characterized by expected variations in the various weather variables such as temperature, precipitation, humidity, cloud cover, and wind speed (Harvey 2000).

There is scientific consensus that human activities over the past several hundred years have increased the emission of *greenhouse gases* (GHGs) including carbon dioxide, ozone, methane and nitrous oxide (IPCC 1990, 1992, 2001; Miller 2002; SRES 2000). These gases absorb the long-wave radiation, or heat, emitted from the Earth's surface, which results in the raising of surface temperatures. While GHGs occur naturally, commencing with the industrial revolution, human activities have contributed to an escalation in the atmospheric concentrations of these gases. It has now become widely accepted that such concentration increases have created an enhanced *greenhouse effect* that has led to (and will continue to lead to) a material change in the global climate system.

The concern over the buildup of GHGs in the atmosphere has been based upon a careful, cause-and-effect assessment of the sequence of events involving emissions of GHGs, their build-up in the atmosphere, the climatic responses to these projected build-ups, and the anticipated impacts of the likely range of climatic responses (Harvey 2000). The significant scientific concern is that the climatic changes will be extremely rapid, unprecedented in human history, and irreversible for all practical purposes (Harvey 2000; Miller 2002). The human activities that contribute to greenhouse gases result from a wide spectrum of actions and decisions taken by individuals, industry and governments. For example, individuals affect GHGs by the type of automobiles that they purchase and the trips that they take, industries by their products and manufacturing processes, and governments by decisions on transportation and energy systems.

Climatic changes can have significant implications for weather patterns at the regional level by influencing not only the general trends in specific weather variables, but also by increasing the incidences of various extreme weather events such as rain and ice storms, heat and cold waves, and hurricanes. Since climate and weather create the conditions under which the natural environment exists, any shifts in their patterns can be expected to concurrently impact a wide array of interrelated biophysical, social and economic areas. Examples of these types of climate-related interactions include: (i) hydrologic cycle and ecosystems including water quality, river flows, groundwater, shorelines, vegetation, wildlife habitat and ecosystems; (ii) human safety due to floods and landslides; (iii) human health, including illnesses, allergies, and responses to extreme temperatures, due for example to changes in air quality and spread of diseases; (iv) commercial and human activities, such as forestry, agriculture, fisheries, tourism, recreation, power production and municipal infrastructures; and (v) social and economic conditions such as jobs and municipal services.

Various regional, national and international groups of scientists and policy makers have been studying climate change and its effects, and how to mitigate and adapt to these changes (CICS 2003; IPCC 1990, 1992, 2001; Miller 2002; SRES 2000). This work can be broadly divided into either scientific studies on climate, e.g. climate models, or studies on the effects of climate change and the requisite adaptation strategies. With passage of the Kyoto Protocol, there have also been strategies developed to reduce anthropogenic GHG emissions (Canada 2002). Strategic environmental assessments (SEA) may be a useful way in which to formulate and assess alternative policies and programs for addressing climate change at the regional or national level and this is especially important for addressing the cumulative impacts related to climate change. For example, an SEA could be undertaken at an overarching level of Canada's overall climate change plan for reducing GHG emissions, while an SEA could also be employed at the sectoral level to address GHG emissions from a transportation activities plan within the Greater Toronto Area. However, the vast majority of environmental assessments that are carried out under federal and provincial legislation are focused at the project level. Hence, the focus of this research is specifically directed at methods for improving the consideration of climate change uncertainties in EAs at the project level.

In EAs, planning for climate change at the project level requires consideration of both a means to adapt to climate change through the design of the project and a means to mitigate climate change through the decision made or choice of alternative selected for the project. The Canadian Environmental Assessment Act (sec. 16) requires that "every screening or comprehensive study of a project…shall include a consideration of…the environmental effects of the project…and any cumulative environmental effects…, the significance of the effects…" The Act (sec. 2) also defines "environmental effects" as "any change that the project may cause in the environment, including any effect of any such change on health and socio-economic conditions…and any change to the project that may be caused by the environment" and defines "environment" to include "land, water and air, including all layers of the atmosphere." Provincial environmental assessment statutes, such as the Ontario Environmental Assessment Act, have similar requirements.

For projects that may contribute, individually or cumulatively, to climate change through the production of greenhouse gases, it is therefore a legislated requirement that these contributions be explicitly considered within the EA. Furthermore, the effects of climate change on the project itself must also be considered since these, too, are "environmental effects". Finally, changes that the project may cause within the environment must be examined in relation to the state of the environment without the project, and these changes may be affected by climate change. Therefore the environmental effects related to climate change that must be considered at the project EA level can be divided into three categories: (i) the contribution/reduction of the project to GHG emissions; (ii) the effects of climate change on the project; and (iii) the effects of climate change on the impacts resulting from the project.

1.3 The Contribution/Reduction of the Project to Greenhouse Gas Emissions

There are a number of projects types for which EAs are carried out that directly contribute to increased GHG emissions and, therefore, to climate change. Examples of these types of project EAs include those: for power plants and other facilities which employ some form of fossil fuel combustion process; and for waste landfills which generate methane. There are also some project types that indirectly contribute to GHG emissions. Examples of these types of project EAs include: any new highway construction that induces a corresponding increase to GHG-emitting vehicular traffic; and expansions to urban development that destroy trees and thereby reduce *carbon sinks*. The direct and indirect increased emissions resulting from such projects have to be both estimated and considered as adverse impacts in environmental assessments. However, determining the degree to which the indirect impacts should considered within the EA process poses a difficult scoping problem and being able to determine the appropriate extent of the analysis of indirect effects remains an open question.

There are also project types that affect GHG emissions but in the opposite direction to those considered above. These projects actually lessen the overall emissions of GHGs, generally by reducing emissions from other processes. Examples of these types of project EAs include those: for new rail and transit projects that reduce road and highway use, and; for hydroelectric and nuclear power plants that replace production from fossil fuel plants. Any estimated reductions in GHG emissions provide a positive impact from the project that need to be factored into the decision-making process during the EA. However, a complete life-cycle analysis would have to be conducted in order to determine the full impact of the project on GHG emissions.

In addition to the emission of GHGs, projects can also create an impact on the local climate conditions. Several examples of these project types are: the effects of a structure on the local microclimate by influencing the wind patterns; the effects of paving natural areas on local temperatures, and; the effects of a new reservoir on local temperatures and humidity conditions. The consideration of these types of local effects has become a well-established component of EA practice and is therefore outside the scope of this research (except with respect to the effects of climate change on impact prediction discussed in a later section).

Each specific project alone can generally make only a minor contribution to GHG emissions when compared with the total emissions on a regional, national or global scale. It is therefore understandable that some proponents would feel that they can either ignore the effect of these emissions on climate change or claim that the resulting effects would be insignificant. While acknowledging that the emissions may contribute to climate change, proponents might argue that this is a larger issue that should be addressed through other means such as government policies, plans and programs. Some stakeholders, however, may want the project's implications for climate change (including the effects of climate change) to be fully addressed in the project EA, noting that when considered together, many such projects may have a significant cumulative effect on climate change, or on Canada's contribution toward meeting the reduction targets set in the Kyoto Protocol. An intermediate view would be for the proponent and stakeholders to try to work together to agree on the scope of the EA.

1.4 The Effects of Climate Change on the Project

Changes in climate and regional weather patterns, including changes in the magnitude and patterns of temperature, precipitation and wind, may affect many types of projects and must therefore be considered during the EA. For example: the design of a marina would be affected by changes in water levels and shoreline; the design of a northern pipeline would be affected by a change in the permafrost; timber extraction on winter roads over frozen lakes would be affected by reduced ice thickness; hydroelectric power production would be affected by a change in streamflows; the design of a building's roof would be affected by a change in snow loads; and the size of a community's stormwater collection system would be affected by a change in extreme rainfall events.

Project designs are generally based upon criteria and standards that have been established using historical data. For example, while building codes for roofs have been based on historic snow loads in the area and the size of stormwater collection pipes has been based on historic rainfall data, these standards may no longer be appropriate under changing climatic conditions. The relative importance of various decision criteria may also be affected. For example, the design of a hydroelectric plant may have been determined on the basis of both power production and flood control using historic streamflow data. With an increase in extreme streamflow events and the risk of flooding, greater importance may need to be placed on future flood control. Finally, the choice of preferred alternative could also be affected by climate change. For example, if the risk of leakage from a northern oil pipeline increases due to changes in the permafrost, another means of transporting the oil (such as tankers) may become the preferred option. The major question for project EAs is to determine how a changing climate might affect the project and how to effectively incorporate the various climate change uncertainties into this decision process.

1.5 The Effects of Climate Change on the Impacts Resulting from the Project

The impacts of a project, such as the effects of a quarry on groundwater levels, the effects of a housing project on a nearby wetland, or the effects of a dam on downstream fisheries, are based on a comparison of the conditions of these environmental components (groundwater, wetland and fisheries) both with and without the project. This is illustrated by the difference between the two solid lines in Figure 1.1, where the horizontal time axis allows for temporal changes. However, any changes in climatic conditions can also cause changes in the conditions of the environmental components. For example, climatic changes may cause a reduction in the groundwater levels, wetland functions and fisheries. The future environmental condition with climate change and without the project is represented by the dashed line in Figure 1.1. The impact of the project relative to this condition, as shown by the shorter arrow, may therefore be significantly different from those predicted without consideration of the effects of climate change.

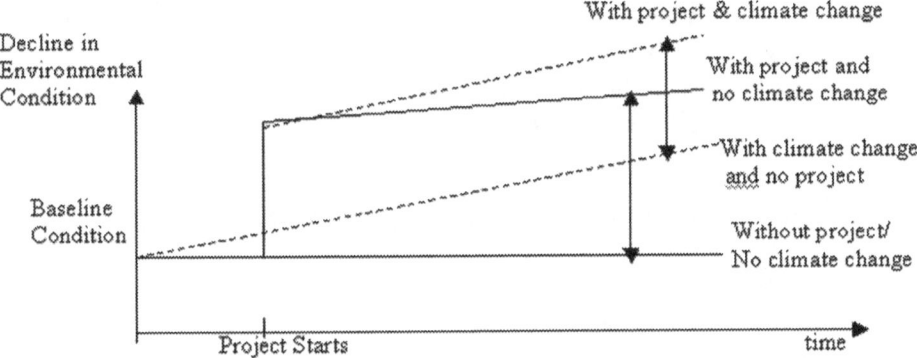

Figure 1.1: Environmental conditions with and without project

Expanding on the fishery example, the impacts of a dam on the downstream fishery would generally based on an assumption that the future condition of the fishery without the project would be the same as its current condition. However, climate change could conceivably materially impact the fishery (i.e. it could entirely disappear due to climate change if such changes significantly affected the local ecosystem), making the impacts of the project relative to this future condition very different.

1.6 Cumulative Effects

The effects of climate change on a project and on a project's impacts may be affected by other current or future projects. For example, the future condition of the fishery discussed above could be affected by other projects in addition to climate change. While these cumulative effects from other projects should be considered in an EA, the focus of this study is solely on the effects of climate change with respect to an individual project.

1.7 Mitigation and Adaptation

When projects contain climate change impacts, the EA process should be used to identify appropriate responses either to mitigate against any GHG emissions or to adapt to any of the effects from climate change. While various agencies and groups have been addressing such mitigation and adaptation responses at the strategic level, similar efforts are also requisite at the project level. However, such project response efforts are outside the scope of this study except for the mitigation and adaptation responses to the *uncertainties* about climate change as discussed in a later section.

1.8 The Uncertainties about Climate Change

The main focus of this study is to address *uncertainties* about climate change in project EAs. While there is broad agreement among climate scientists about the mechanisms of, and general trends in, global warming, there are significant uncertainties about some key aspects of climate change and its effects (Harvey 2000; Miller 2002). These uncertainties include: the average global temperature increase that would result from specific increases in greenhouse gases; the effects on regional climate variables such as temperatures, precipitation and wind, and their fluctuations; the effects of projected climate change on systems such as agriculture, oceans, ice sheets and terrestrial ecosystems; the spread of diseases and other related hazards to human health; and the resulting effects on socio-economic systems such as jobs and health care.

In general, the most needed climate information required for project-specific EAs concerns the specific climate for the region in which the project is being undertaken. Unfortunately, any determination of specific predictions at the regional level of detail requires the production of climatic information in which there is the least degree of confidence. Furthermore, there are likely to be many highly uncertain future relationships between climate change and various other biophysical, social and economic impacts. Using the previous fishery example, there may be significant uncertainties about the effect of climate change on the fishery both with and without any proposed upstream dam.

Figure 1.2 illustrates the effect of climate change on the prediction of project impacts. Uncertainties about the "without project" and "with project" environmental conditions are shown as bands, and the differences between them represent widely uncertain predictions about the effects of the project.

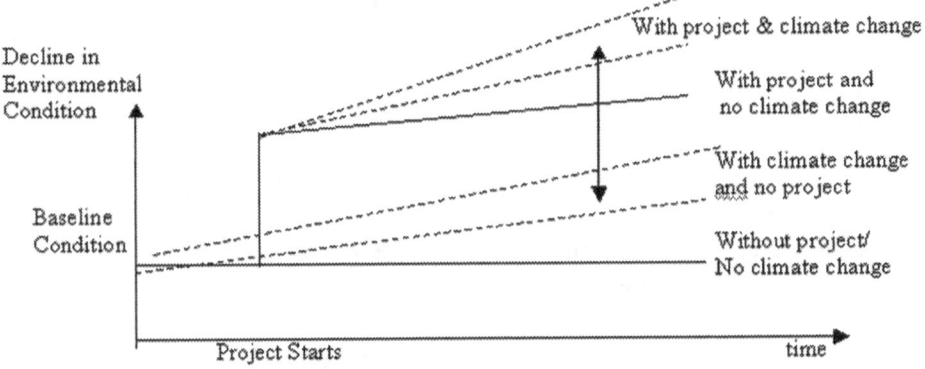

Figure 1.2: Effect of uncertainties on impact prediction

As the relationship between climate change and any potentially affected environmental components becomes less and less direct, the uncertainties can become more and more pronounced. Using an example where climate change will affect precipitation in a watershed and water in the river is used for irrigation purposes, the relationships between rainfall and streamflow, streamflow and irrigation, irrigation and agricultural production, agricultural production and farm jobs, and farm jobs and the local economy are all uncertain to varying extents. There are also the cumulative effects of these chains of uncertainties, which would make the prediction of the impacts of climate change on the regional economy highly uncertain. Irrespective of these uncertainties, decisions will have to be taken and it proves to be a better course of action to adopt the *precautionary principle* when making these decisions by recognizing and addressing these inherent uncertainties rather than by ignoring them.

1.9 Environmental Assessment Practices Related to Climate Change

In a related research project, Murphy (2003) reviewed eleven recent EA reports to assess the state of EA practices concerning climate change and some findings from this study are briefly summarized in this section. Only assessments performed after 1988 were considered, since this corresponded to the time when the issues of accelerated anthropogenic climate change were elevated into the global forum and when the Intergovernmental Panel on Climate Change (IPCC) was established. While climate change and its uncertainties held direct relevance to each of the eleven projects, the study examined whether and how these climate issues had been addressed within their EAs. A summary of these findings follows in the remainder of this section.

Five of the EAs were chosen because they had been reviewed by the Canadian Institute for Climate Studies (CICS) at the University of Victoria as part of a recent research project on climate change and EAs funded by the Canadian Environmental Assessment Agency (CEAA) (Lee 2001). The projects evaluated in these EAs, including the dates of the assessments, were for the: Diavik Diamond Mine (1998)–C(U), E; Northumberland Bridge Crossing (1990)–C(U), E(U), P(U); Cascade Heritage Park Hydroelectric Project (1996)–C, E; Little Bow Hydroelectric Project (1995)–N and, ; Elliot Lake Mine Decommissioning (1995)–E(U) [Note: the C, U, P, E, N notation will be explained subsequently]. The remaining six EAs were selected based upon recommendations from EA practitioners at Environment Canada (Murphy 2003) and included the: Mattagami River Hydroelectric Generating Station Extensions (1990)–N; Little Jackfish Hydroelectric Project (1992)–C; White Rose Offshore Oil and Gas Project (2000)–C, E; Aquarius Gold Mine Project (1999)–N; St. Mary's Dock Expansion (1990)–N, and; Pickering Nuclear Generator A Re-Start (2000)–N. The Mattagami River and Little Jackfish hydroelectric project EAs were done under the Ontario EA Act process, while the other nine were conducted under the Canadian EA Act or the Federal EA Review Process.

Each EA was reviewed to determine both whether and how: (i) climate change had been considered; (ii) uncertainties about climate change had been considered, and; (iii) climate change affected decisions on the project. The review demonstrated that the five EAs identified by the "N" notation gave no consideration to climate change, even though the projects contained issues that were directly related to climate change. For example, the two hydroelectric projects would reduce GHG emissions elsewhere and would be affected by changes in streamflows; the gold mine could significantly affect and be affected by the local hydrology; the dock expansion would be affected by changes in water levels and extreme weather events; and the nuclear generating station would reduce GHG emissions elsewhere. However, it should be noted that following the submission of these EAs, there may have been some consideration of climate change by the proponents or others involved.

The remaining six EAs incorporated some degree of consideration for climate change. The nature of these considerations was categorized according to the project's: (i) contribution to, or lessening of, GHG emissions, identified by the "C" above; (ii) effects of climate change on the project, identified by "E"; and (ii) effects of climate change on the project impacts, identified by "P". It was also determined whether climate change uncertainties had been addressed in some way within each of the three categories. A "(U)" next to a C, E, or P identifies the projects in which uncertainties had been considered.

Below is a summary of how climate change had been addressed in the EAs of the six projects. (1) The Diavik Diamond Mine EA evaluated GHG emissions from the project, incorporated energy efficiency into the design to reduce emissions, and considered the effect of climate change on permafrost. (2) The Northumberland Bridge EA addressed the delay in ice-out and rising sea levels as a result of climate change, considered these in the bridge design, and indicated that the cumulative effects of the project and climate change on fisheries were not known. Climate change also played a significant role in the report of the Review Panel. (3) The Cascade Heritage Park Hydroelectric EA addressed GHG production that would result from flooding of the shoreline vegetation, considered that the project would replace energy produced by fossil fuels, and recognized that hydrologic variations could decrease energy production. (4) The Elliot Lake Mine Decommissioning EA considered extreme weather events, addressed increased evapotranspiration and drought from climate change, and the Panel report determined climate change to be a concern.

(5) The Little Jackfish Hydroelectric EA recognized that reservoir creation could affect the microclimate, e.g. wind patterns, air temperatures, etc. (6) The White Rose Oil and Gas EA estimated the project's GHG emissions and considered the effect of climate change on ice coverage.

The outcome from this review of the eleven EAs showed that climate change had not been adequately acknowledged or addressed across environmental assessments, that uncertainties about climate change have been addressed even less well, that climate change had been addressed inconsistently across similar types of projects, that more recent EAs are not necessarily better with respect to climatic concerns, and that some responsible authorities and panels have raised climate change as an important issue.

The CICS review of selected EAs mentioned above included the following key findings related specifically to the uncertainties resulting from climate change: the projects used a range of techniques to measure the sensitivity of the project to climate change; many EA proponents felt that scientific knowledge about future climate was too uncertain to have the confidence to act upon; in cases where climate change information was considered to be inadequate for decision making, the need for contingency plans and environmental monitoring was recognized; climate change uncertainty must be addressed in EAs and a range of probable futures made available to support sensitivity studies in relevant projects, and; that there is a need for a formal guide that should, among others things, present information on the reliability and uncertainty of projections.

1.10 Conclusion

Climate change represents an essentially irrevocable environmental phenomenon that is global in scope. The consequences of a changing climate have significant implications for projects that affect the environment and consequently must receive appropriate consideration in EAs. However, climate change impacts have been inadequately considered in project EAs and the corresponding uncertainties have been addressed even more poorly. To redress several of the identified shortcomings, this project focuses upon methods that can be undertaken within EAs for addressing and communicating the numerous uncertainties resulting from climate change.

Incorporating climate uncertainties into EAs is neither straightforward nor easily accomplished. Three basic techniques will be presented for incorporating climate change impacts and uncertainties into EAs: scenario analysis, sensitivity analysis, and probabilistic analysis. While scenario analysis has been the approach most often associated with climate change studies, sensitivity analysis and probabilistic analysis represent more general utilitarian techniques that have been widely applied in the analysis of uncertainties. To evaluate project uncertainties and project sensitivities under a range of potential climate futures, these three methods will be used to examine the outcomes of both climate change impacts on a project and project impacts on climate change. Once these impacts have been analyzed, it is essential that the identified uncertainties be clearly communicated to the diverse body stakeholders involved in the EA process.

Several recommendations for effectively accomplishing all of these analysis and communication tasks will be developed throughout the course of this study and a significant set of examples based upon a hydroelectric project will be used to articulate the application of the various concepts and methods employed. Proponents would obviously be required to adapt and modify any of these recommendations to fit the very specific situations faced in their own EAs. In summary, the overarching objective of the study is to advance the practice of EAs within Canada through the inclusion of an appropriate consideration of climate change issues.

The remaining chapters in this study are organized according to the following topic sequence. Chapter 2 provides a background review of the causes, impacts, and uncertainties of climate change and also discusses recently adopted national and international climate initiatives. In chapter 3, the three basic analytical approaches for addressing and analyzing climate uncertainties are presented. The hydroelectric example is introduced in Chapter 4 to illustrate the types of impacts commonly considered in project EAs and to examine which of the analytical methods, or combinations thereof, prove most appropriate for addressing the specific impacts. Chapter 5 reviews numerous possible methods for communicating climate uncertainties to stakeholders possessing differing levels of technical sophistication. Finally, chapter 6 concludes with a comprehensive synopsis of the numerous suggestions and recommendations for effectively incorporating, analyzing, and communicating the uncertainties of climate change in project EAs.

2

Climate Change Causes, Impacts and Uncertainties

2.1 Introduction

Anthropogenic global warming, or climate change, has become one of the most visible, far-reaching, and studied global environmental issues of the past two decades. To the layperson, global warming is a term synonymous with hotter weather rather than with its more profound and unfamiliar environmental consequences (Kempton 1991). *Weather* can be defined as the day-to-day changes in the state of the atmosphere at a specific location that includes variables such as temperature, humidity, wind, cloud cover, and precipitation (Harvey 2000). *Climate*, however, can be defined in part as the average weather, in which a regional climate is further characterized by its expected year-to-year variations in the various weather variables (Harvey 2000). While two regions can possess the same mean annual temperature, if one region experiences greater annual variations in its weather patterns, then these regions would be considered to possess different climates. The *climate system* of a given region is formed by the interaction of all of its components acting together and consists of much more than just the atmospheric conditions. When climate scientists refer to the concept of global warming, they are referring not only to atmospheric conditions, but also to multiple ecosystem and geophysical system effects in which the most significant impacts can include the extinction of numerous animal species and major range shifts in various natural ecosystems.

There is scientific consensus that human activities over the past 200 years have created an enhanced greenhouse effect which has (and will continue to) lead to a significantly warmer global climate system (IPCC 1990, 1992, 2001; SRES 2000). The main basis for concern regarding the greenhouse effect is not centred upon the already observed temperature increases. Rather the concern is focused upon the dramatic increases to the anthropogenically generated greenhouse gas emissions that have already occurred, the prospect of massively greater increases in these emissions, the strong theoretical predictions that such increases will soon elicit a massive climatic response, and that, once initiated, these changes will be irreversible for all practical purposes (Harvey 2000). The major policy issue is to anticipate what the impacts of these substantially larger greenhouse gas increases will be and to determine how to limit the increases should these impacts be considered undesirable.

2.2 Causes of Climate Change

All objects emit energy in the form of electromagnetic radiation. The absorption of any radiation by an object contributes to its warming, while the emission of radiation by the object contributes to its cooling. There are certain gases that exist within the atmosphere that trap energy through the absorption of infrared radiation, thereby leading to a warmer climate. This process has commonly been referred to as the *greenhouse effect* and the contributing gases are known collectively as *greenhouse gases* (GHGs). The greenhouse effect occurs naturally and the global climate would be approximately 33°C cooler in its absence. On the other hand, GHGs can occur both naturally and artificially. The naturally occurring gases include water vapour (H_2O), carbon dioxide (CO_2), ozone (O_3), methane (CH_4) and nitrous oxide (N_2O), while the main artificial gases are chlorofluorocarbons (CFCs), hydrochlorofluorocarbons (HCFCs), hydrofluorocarbons (HFCs), and sulfur hexafluoride (SF_6). While water vapour is the most influential GHG, human activities have magnified the greenhouse effect by directly emitting certain GHGs (CO_2, CH_4, N_2O, CFCs, HCFCs, HFCs, SF_6) into the atmosphere and by indirectly emitting certain other non-greenhouse gases (carbon monoxide–CO, nitrogen oxides–NO_X, sulfur oxides–SO_X, hydrocarbons) which react chemically with the atmosphere to either create additional surface-level O_3 (i.e. smog) or extend the lifetime of other GHGs (CH_4, HCFCs, HFCs).

The atmospheric concentrations of GHGs have increased dramatically since the start of the modern industrial era (approximately 1800 AD) and these increases can clearly be attributed to human activities. Over the past 160,000 years, CO_2 has undergone natural variations in concentration ranging from 180 ppmv (parts per million by volume) up to 300 ppmv, while CH_4 concentrations have varied between 0.3 ppmv and 0.7 ppmv (Harvey 2000). In contrast, over the past 200 years, the CO_2 concentration has increased to over 360 ppmv and the CH_4 concentration currently stands at over 1.7 ppmv. These concentrations are now higher than at any time in the last 420,000 years and the concentrations of all of the other GHGs have demonstrated similarly abrupt increases (Miller 2002).

The unprecedented rates of these observed changes, therefore, far exceed anything that has ever occurred naturally and there is essentially no debate within the scientific community concerning the anthropogenic origin of its causes (Harvey 2000; Miller 2002; IPCC 2001). Independent, supporting lines of evidence indicate that: the rate of increase in atmospheric GHG concentrations over the past century matches the rate of anthropogenic emissions; the atmospheric oxygen content has been declining at the same rate as fossil fuel emissions of CO_2 have been increasing (oxygen is consumed whenever fossil fuels are burned), and; the changes in atmospheric proportions of carbon isotopes indicate that the atmosphere is becoming enriched with carbon from fossil fuel sources rather than from natural sources.

Until the industrial revolution, the earth's natural carbon cycle remained essentially in equilibrium, with vast quantities of carbon continually circulating throughout the planet's atmosphere, oceans, and terrestrial biosphere. Scientific evidence indicates that the undisturbed terrestrial biosphere and the oceans, which are the only possible alternative candidates available to explain the currently observed CO_2 buildup, have both been sinks rather than sources of atmospheric CO_2. Furthermore, the Intergovernmental Panel on Climate Change (IPCC) have estimated that 70 to 90 per cent of the increase in CO_2 emissions is attributable to fossil fuel burning with the remainder due to land-use changes, particularly deforestation (IPCC 2001). Hence, while CO_2 and other GHGs have both natural and human sources, it is the anthropogenic emissions that are responsible for the increase in atmospheric concentrations.

On the basis of verifiable scientific measurements, these GHG concentration increases lead to a significant trapping of heat and, on the basis of fundamental physical principles and observational evidence, this additional heat trapping will lead to changes in the earth's climate which will be significant from both a human and an ecological perspective (Harvey 2000; IPCC 2001). Climate changes in the future will depend explicitly upon (i) the future emissions of GHGs, (ii) the buildup of GHG concentrations from these emissions together with their resultant heat trapping, and (iii) the *climate sensitivity* (or temperature change) for a given increase in GHG concentrations. Scientists have already concluded that over the last century, the global average temperature of the earth's surface has risen by 0.6°C (Harvey 2000; IPCC 1990, 1992, 2001; Miller 2002). However, under a *business-as-usual* (BAU) projection for the 21st century in which there is no effort to restrain GHG emissions, the climate will become as warm as at any time during the past 125,000 years under the lowest temperature (or low climatic sensitivity) estimate and warmer than at any time during the previous 38 million years for the highest temperature (or high climatic sensitivity) estimate (IPCC 2001, Harvey 2000, Miller 2002).

Hence, the concern over the buildup of GHGs in the atmosphere has not been based upon a simple extrapolation of past trends in concentrations or temperatures. Rather, it has been based upon a careful, cause-and-effect assessment of the sequence of events involving emissions of GHGs, their build-up in the atmosphere, the climatic responses to these projected build-ups, and the anticipated impacts of the likely range of climatic responses (Harvey 2000). The basis expressed for the significant scientific concern is that the climatic changes foreseen will be extremely rapid, unprecedented in human history, and irreversible for all practical purposes.

2.3 Impacts and Uncertainties of Climate Change

The extra heat trapping of infrared radiation arising from increases in the GHG concentrations is commonly referred to as a *radiative forcing*. The forcings for each GHG are well-established, since they can be explicitly determined from laboratory measurement. A doubling in the concentration of CO_2 (or "$2xCO_2$") traps an additional 4 W/m^2 of infrared heat when averaged over the entire globe (Harvey 2000). Although gases other than CO_2 have been increasing in concentration, the overall climatic response depends largely on the net radiative heating of all GHGs acting together, not on the individual gases responsible for the heating. Hence, 2x CO_2 has been used as a surrogate for any combination of GHG increases which is radiatively equivalent to a globally averaged heat trapping of 4 W/m^2. If everything except the temperature remained constant following an extra heating of 4 W/m^2, then the earth's climate would warm by 1.2 °C. The exact magnitude of this temperature increase has been determined on the basis of sound, verifiable, and validated scientific principles (Harvey 2000).

However, once the earth's surface temperatures begin to increase, other conditions, or *feedbacks*, start to occur which further increase the trapping of infrared radiation as well as increasing the absorption of solar energy. These radiative feedbacks lead to additional temperature changes in an effort to maintain the requisite equilibrium balance between infrared emissions and solar energy absorption. The three major radiative feedback processes identified in the climatic response to an increase in the earth's temperature (Harvey 2000) are due to: (i) the increase in atmospheric water vapour; (ii) the reduction in the earth's reflectivity, or *albedo*, from the melting of ice and snow, and; (iii) the resulting changes in cloud patterns. The impact from the increased concentration in water vapour is expected to produce the single most significant radiative feedback.

The impacts and consequences of these three feedbacks can be summarized in the following way (see Harvey 2000). Feedback (i): A warmer climate naturally leads to increases in both the evaporation of water and the amount of water vapour that the atmosphere can hold. Since the warmer temperature leads to more water vapour in the atmosphere and since water vapour is a naturally occurring GHG, the net consequence of this increased concentration is for an additional warming of the atmosphere to occur. Based upon evidence from a wide variety of direct observations, the combined impact of the 1.2 °C temperature increase together with the increased concentration of water vapour leads to an overall warming of between 2.0 to 2.5 °C (i.e. a doubling of the initial warming effect). Feedback (ii): A warmer climate naturally produces a general increase in the melting of ice and snow on the earth's surface (with a particularly significant effect in the higher latitudes). This melting reduces the earth's albedo leading to the absorption of more solar energy, thereby leading to an increase in the earth's temperature of between 0.0 to 0.5 °C. Thus, in combination with the initial temperature change and feedback (i), the melting of ice and snow leads to an overall warming of between 2.0 to 3.0 °C. Feedback (iii): Due to a variety of atmospheric factors, a warmer climate leads to natural changes in the patterns of cloud cover. Unlike feedbacks (i) and (ii), clouds produce two competing climatic effects. On one hand, they produce a warming effect by trapping infrared radiation, while on the other hand, they provide a cooling effect by increasing the planetary albedo by reflecting solar radiation. The net impact of the overall cloud feedback effect depends explicitly upon where the clouds occur on the earth, whether the clouds occur during the day or the night, and how high the clouds are. Because of the complexity involved in estimating these competing cloud cover impacts, the net effect of cloud changes could produce either a positive, warming feedback or a negative, cooling feedback. However, the best scientific estimates for the temperature impact from the cloud feedback in combination with both the 1.2°C temperature increase and feedbacks (i) and (ii), is for an overall globally average equilibrium warming of between 1.5 and 4.5 °C (Harvey 2000; Miller 2002; IPCC 2001).

The equilibrium globally averaged warming for $2xCO_2$ is generally referred to as the *climate sensitivity*. Climate sensitivity is the single most important aspect of global warming, since this is the key feature driving every other response. The actual climate sensitivity depends explicitly upon the climate feedbacks and the more positive the climate feedbacks, the greater the climate sensitivity. Due to variations in the way in which different models have treated the various feedbacks, climate scientists have estimated that the climate sensitivity lies somewhere within the range 1.5 to 4.5 °C (IPCC 2001). As described above, this factor of three uncertainty can be largely attributed to the uncertainty in the net feedback effect of clouds (Harvey 2000; IPCC 2001) and it is extremely unlikely that this uncertainty range can be reduced at any point in the near future.

In addition to the principal uncertainty regarding the actual value of climate sensitivity, there are a number of other key uncertainties surrounding a changed climate (Miller 2002). There are major uncertainties concerning the impacts of projected climate change on various systems such as agriculture, forestry, oceans, and non-forest terrestrial ecosystems. In agriculture, uncertainties exist regarding the extent to which adaptation can mitigate the adverse impacts of climate change and concerning the benefits of CO_2 fertilization (Wolfe & Erikson 1993). In forestry, there is uncertainty concerning how effectively forests will be capable of migrating, the impacts of CO_2 fertilization, and whether current predictions overestimate the initial impacts of climate change on the forests (Morgan *et al.* 2001; Steffen 2001; Loehle 1996; Taubes 1995). There is uncertainty regarding how large a warming would be required to provoke the destabilization of the West Antarctic ice sheet and an irreversible melting of the Greenland ice sheet; it could occur with as little as a 2°C warming (Harvey 2000; IPCC 2001; Miller 2002). There is uncertainty regarding the projected sea level rise during the next century (25-70 cm) and with the rise over the longer term (10-13 m) (Harvey 2000). There is uncertainty with the amount of warming required to provoke a massive dieback of the world's coral reefs; it could occur with as little as a 1°C warming (Kleypas & Opdyke 1998; Normill 2000; Showstack 2000; Wellington *et al.* 2001). There is uncertainty regarding the extent of species extinctions (Henderson 1989; Pounds 2001), the likely spread of diseases into new regions (Stone 1995), and other related hazards to human health (Epstein 2000; Roberts 1988).

There is also uncertainty regarding possible feedbacks between climate and the carbon cycle, in which the initial warming provokes a significant pulse of CO_2 into the atmosphere through dieback of forests and respiration of soil carbon, thereby leading to a further warming and a further CO_2 flux into the atmosphere. This uncertainty is related to the uncertainty concerning the present role that the terrestrial biosphere plays as a carbon sink and, hence, concerning how large the stimulatory effect of a higher atmospheric CO_2 concentration has to be on ecosystem productivity in order to match the observed rate of increase in atmospheric CO_2 (IPCC 2001; Miller 2002). Although such a positive feedback would not lead to an out-of-control greenhouse effect, it could nevertheless significantly increase the combined human CO_2 emissions over the course of a century.

For specific individual regions, there are significant uncertainties concerning both temperature and precipitation changes. It is these regional changes, together with their seasonal distributions, which will determine the actual climatic impacts, not the globally averaged effect. For example, in the interior of North America, summer warming can be expected to range from as little as 2 °C to in excess of 8 °C. This temperature change will be accompanied by precipitation increases in some parts of the continent and precipitation decreases elsewhere including areas subject to the greatest increase in warming (Miller 2002). Thus, the actual climate information most needed for specific planning, the regional detail, is the information in which there is the least confidence (Harvey 2000).

However, in spite of all the inherent uncertainties, there is broad agreement among climate scientists regarding certain climate change impacts that can be expected in response to a CO_2 doubling. The features in which there is very high confidence are that: (i) the largest warming will occur at high latitudes due to the melting of ice and snow, (ii) greater warming will occur during the winter at high latitudes, but greater warming will occur during the summer where soils become drier, (iii) there will be a greater warming of the continents than the oceans except at very high latitudes, (iv) there will be an overall increase in the intensity of the hydrological cycle, since warmer oceans will undergo greater evaporation, which must be balanced by increased precipitation, and (v) the stratosphere will cool. There is less confidence in projections that: (i) the midlatitude precipitation belts will shift poleward, (ii) soils will become drier in summer at many mid-continent locations, (iii) the variability of precipitation will increase in many areas with an increase in the frequency of droughts and floods, (iv) there will be a greater fraction of rain falling as intense, convective rainfall such that more will be lost as runoff, and (v) there will be a tendency for the Asian and West African monsoons to become stronger.

2.4 Recent International Initiatives Adopted to Address Climate Change

In 1992, the United Nations Framework Convention on Climate Change (UNFCC) was signed and ratified by 186 countries in Rio de Janeiro. The legally-binding and declared objective of the UNFCC can be stated as follows. "The ultimate objective of this Convention and any related legal instruments that the Conference of Parties may adopt is to achieve, in accordance with the relevant provisions of the Convention, *stabilization of greenhouse gas concentrations in the atmosphere at a level that would prevent dangerous anthropogenic interferences with the climate system* [emphasis added]. Such a level should be achieved within a time frame sufficient to allow ecosystems to adapt naturally to climate change, to ensure that food production is not threatened and to enable economic development in a sustainable manner." As described in the earlier sections, considerable uncertainty exists with regard to the climate sensitivity and the resulting consequences of increased concentrations of anthropogenic GHGs. While the minimum estimated climate sensitivity for a CO_2 doubling of 1.5 °C might not be particularly alarming, if the climate sensitivity of 4.5 °C occurred, there would be an extremely high risk of catastrophic climatic consequences. Since the stated goal of the UNFCC is to avoid dangerous anthropogenic interference in the climatic system, it follows that the current national/international policy of the signatory countries must be directed toward a stabilization of GHG emissions. Thus, the UNFCC requires that the signatory countries must address GHG emissions in the face of the impacts from these inherent uncertainties.

In 1997, the Kyoto Protocol covering such GHGs as CO_2, CH_4, N_2O, HCFCs, HFCs, SF_6 was signed. Under this protocol, the Annex I, or developed, countries are required to reduce their CO_2-equivalent GHG emissions by 5 to 8% from the 1990 levels by 2008-2012. Given that emissions would otherwise increase by 20^+%, the protocol provides a significant first step in the control of GHGs. In 2002, Canada ratified the protocol thereby committing the country to achieving the requisite emissions reductions (Canada 2002). During both the pre- and post-ratification stages, numerous business and political groups within the country have espoused vehement opposition to the protocol, generally citing the major uncertainties in the risks that might arise from climate change and in the unknown consequences to be incurred in reducing anthropogenic GHG emissions (Mackie 2002; Miller 2002; Toronto Star 2002a, 2002b, 2002c, 2002d). However, the net effect of this ratification is that, in spite of, or because of, the many uncertainties surrounding climate change, organizations operating within Canada now have fiduciary obligations to address the potential consequences of their activities on climate change under the conditions of both Kyoto and the UNFCC.

2.5 Conclusion

Uncertainty has often been cited as one of the main reasons for delaying actions to reduce GHG emissions. However, this position is contrary to the principles of sound risk management, since a proper consideration of the impact of uncertainty increases, rather than decreases, the rationale for preventative action. This *precautionary principle* holds especially true when these impacts are irreversible. None of the uncertainties can override the fact that GHG emissions will need to be strongly and rapidly restrained, both within Canada and globally. The effects of the uncertainties concern the extent of damage that the earth's climate is already committed to as the climate "catches up" to the GHG build-up that has already occurred and the additional damage associated with the various future GHG concentration increases (Miller 2002). Therefore, climate change requires immediate attention because of these uncertainties, not in spite of them.

3

Methods for Addressing Climate Change Uncertainties

3.1 Introduction

In this chapter, three basic approaches are presented for addressing and analyzing the uncertainties arising from climate change studies: *scenario analysis*, *sensitivity analysis*, and *probabilistic analysis*. The approach most often associated with climate change studies is scenario analysis, whereas the other two methods, sensitivity analysis and probabilistic analysis, are general techniques more widely used for addressing uncertainties. In the subsequent chapter the use of these three approaches for addressing climate change uncertainties in environmental assessments will be illustrated through an application to a run-of-the-river hydroelectric project in Northern Ontario.

3.2 Scenarios

Greenhouse gases such as water vapour, carbon dioxide, methane and nitrous oxide absorb the long-wave radiation, or heat, emitted from the Earth's surface and the subsequent re-emission of this energy results in raised surface temperatures. While greenhouse gases occur naturally, commencing with the industrial revolution, human activities have contributed to an upward escalation in the atmospheric concentrations of these gases. It has now become widely accepted that such concentration increases have materially affected the global climate. In light of these changes, it has become essential to integrate the potential impacts that might arise from future global climate change into the planning of all large-scale engineering projects. In order to determine how the future climate might change it is imperative to understand how the concentrations of the atmospheric components that impact the Earth's energy balance may change. It is also important to clarify how these greenhouse-induced changes differ from all of the natural climatic variations that already occur (Ausebel 1991). As this would necessitate that assumptions be made concerning how both the natural and anthropogenic emissions of these greenhouse gases will change, the determination of exactly *how* future atmospheric compositions may change is fraught with considerable uncertainty. Furthermore, since greenhouse gas (GHG) emissions are the product of numerous complex dynamic systems driven by forces such as demographic development, socio-economic development, and technological change (SRES 2000), their future evolution is highly uncertain.

The assessment of the impacts from these changes has led to the prevalent tendency to consider entirely different sets of "futures" or *scenarios* as exemplified, in particular, by the series of scenarios constructed by the Intergovernmental Panel on Climate Change (IPCC) (IPCC 1990; 1992; SRES 2000). Scenarios provide alternative images of how the future might unfold (Kirkwood 1997) and are considered an appropriate tool with which to analyze how the different driving forces might influence future emission outcomes and to assess the associated uncertainties. Since these driving forces could follow many possible paths, it is important to examine GHG emissions under alternative sets of self-contained and consistent emissions assumptions. Scenario development involves developing a plausible story or a "scripting" of the future (Kirkwood 1997; Kleiner 1994; Schank 1990). Any scenario necessarily includes subjective elements and is open to various interpretations (Schank 1990). While preferences for different scenarios necessarily vary among users, the IPCC analyses explicitly proffer no judgments as to the preference for any of their scenarios and no probabilities of occurrence are assigned (SRES 2000). The goal of scenario analysis is to design a series of potential future outcomes that do not depend upon hindsight but result from user-defined emissions patterns under a variety of possible operating environments.

Scenario analysis requires the ability to progress beyond the scope of familiar and current contexts in order to address various potential futures (Dawes 1988; Kirkwood 1997; Kleiner 1994; Schank 1990; Schoemaker 1995). While in theory, every possible future emission assumption should be considered, in practice, the number of scenarios analyzed is reduced to a subset that efficiently and reasonably represents their possible ranges. Generating a representative set of scenarios is paramount in any practical implementation of a multi-scenario model since the results depend upon it. Implicit in this approach is the recognition that the future cannot in any way be predicted, and may not resemble what might be expected based on a simple continuation of present-day trends and patterns (Kirkwood 1997). Scenarios provide decision-makers with a long-term context for near-term analysis, even when recognizing the inherent uncertainties in the long-term projections (Kirkwood 1997; Schank 1990; Schwartz 1991). By simultaneously considering multiple scenarios in an analysis, one minimizes the risk that the analysis is merely an opinion about the future (Kirkwood & Seidman 1985; Russo & Schoemaker 1989).

3.2.1 Scenarios Created by the IPCC

The scenarios that have received the most widespread circulation, acceptance and implementation in long-range planning studies have been those created by the IPCC. In the 1992 Supplement (IPCC 1992) to the First Assessment Report (IPCC 1990), Leggett *et al.* (1992) produced the six IS92 emissions scenarios that reflected the large uncertainty associated with the evolution of population and economic growth, technological advances, technology transfer, and responses to environmental, economic or institutional constraints. Until recently, the IS92a *business-as-usual* scenario had been in widespread use by the climate modelling, vulnerability, impacts and adaptation communities. While the IS92a scenario drew considerable attention, the six distinct emissions scenarios were all considered to be equally likely.

In the IPCC Third Assessment Report (IPCC, 2001), a new set of emissions scenarios for use by policy-makers was commissioned to replace the IS92 scenarios. The IPCC Special Report on Emissions Scenarios (SRES 2000) presented this new set of scenarios updating the content of the 1992 scenarios by incorporating a better understanding of the driving forces behind the emissions and methodologies—with the specific inclusion of the carbon intensity of the energy supply, income gaps between developing and developed countries, and the various impacts from sulfur emissions. SRES (2000) used alternative models to create four different storylines of qualitative emissions drivers. The theme behind each storyline represented a different combination of demographic, social, economic, technological and environmental developments. Several international modelling teams collaborated in quantifying these SRES storylines and the resulting emissions scenarios represent the quantitative interpretations of these qualitative storylines.

This newly created set of scenarios represents a diverse range of driving forces and emissions that more accurately reflects current understanding and knowledge of climate change. As with the IS92 scenarios, the possibility for the occurrence of any single emissions path as described in these scenarios is highly uncertain and no single scenario can be treated as more or less probable than others (SRES 2000). The scenarios were considered plausible business-as-usual scenarios, since they did not assume that societies would be taking any deliberate actions to reduce their GHG emissions. The SRES scenarios reinforce the earlier understanding that the main drivers of GHG trajectories will continue to be social and economic development, and the rate and direction of technological change. A significant outcome from the updated scenario families is that similar future GHG emissions can result from very different socio-economic patterns, and similar developments of driving forces can result in very different future emissions (SRES 2000). Furthermore, the convergence of regional per capita incomes can lead to either high or low GHG emissions.

3.2.2 Atmospheric General Circulation Models

To make projections of future patterns of climate change, researchers use three-dimensional computer models known as *atmospheric general circulation models* (AGCMs or GCMs). In these models, physical quantities which vary continuously in three dimensions are represented by their values at a finite number of points arranged in a three dimensional grid. This is necessary because only a finite number of calculations can be performed. The spacing between the gridpoints is the *spatial resolution*. The finer the resolution, the more points, and the more calculations that need to be performed. Hence, the resolution is limited by the computing resources available and the typical AGCM resolution is hundreds of kilometres in the horizontal. The Canadian Climate Impacts Scenarios (CCIS 2003) project provides ready access to climate scenario information, data, and scenario construction advice for researchers from both their own AGCM models, while also providing scenario data from many of the other major international climate institutes.

While there is no single most likely, central, or best-guess scenario with respect to the underlying scenario literature, the distribution of possible outcomes generated by considering various scenarios provides a useful context for understanding the relative position of the likelihood of an occurrence. Hence,

SRES scenarios provide one set of consistent data that could be used in diverse planning studies, although there are many other scenarios that could be examined. In any study that is directly or indirectly influenced by climate change impacts, it is absolutely essential to consider a range of different scenarios with a variety of assumptions regarding the driving forces. Depending upon the nature of the application, the important uncertainties would range from the actual GHG emissions to the specific underlying driving forces that generated the emissions. For example, climate modellers who want to cover a range of cumulative emissions categories to assess the robustness of options in terms of impacts, vulnerability, and adaptation, would select scenarios with similar emissions but very different socio-economic characteristics. In mitigation studies and analysis, scenarios providing variation in both emissions and socio-economic characteristics would be necessary. And for policy analysis at the national scale, the most appropriate scenarios may be those that best reflect specific circumstances and perspectives. The major strength for applying scenario analysis to climate change studies is that they provide a common and consistent basis from which all researchers can work and thereby permit a comparability between different studies.

Given the significant uncertainty regarding the future GHG emissions, scenarios should be considered as part of a comprehensive assessment of the impacts of climate change. The IPCC recommends that "users should design and apply multiple scenarios in impacts assessments, where these scenarios span a range of possible future climates, rather than designing and applying a single 'best guess' scenario" (IPCC 2001). While this IPCC recommendation leads to the question as to which of the many scenarios available should be used, the brief answer is as many as possible but should include scenarios constructed from two different AGCMs at the very least (CCIS 2003). Reviewing several different scenarios can help to produce a broad view about what is likely to happen and how key variables can change (Jones 1991; Kirkwood 1997). Should time prevent the use of anything more than a small number of scenarios, then the scenarios selected should represent the extreme range of changes projected for the region in question, as well as a scenario which reflects a more intermediate degree of change (CCIS 2003).

Although AGCM models are in general agreement concerning the global averaged effects, they differ considerably with respect to the determination of regional details. Due to the resolutions used in AGCMs, climate scientists stress that climate models cannot provide certainty about the details of specific changes in specific places. This is unfortunate since it is the regional changes that determine the specific climatic impacts and many important elements of climate systems (clouds, land surface variations) have scales much smaller than the resolutions used in the AGCMs. Thus, the information most needed, the regional detail, is the information in which there is the least confidence (Harvey 2000). Detailed computer models at fine resolutions are available, but these are computationally too expensive to be included in a climate model, and the AGCMs have to represent the effect of these sub-grid-scale processes on the climate system at their coarse grid-scale. A formulation of a small-scale process on the large scale is called a *parameterization* and all climate models are forced to adopt a parameterization to some extent. A number of methodologies have been developed for deriving more detailed regional and site scenarios of climate change for regional impacts studies. These downscaling techniques are generally based upon AGCM output and have been designed to bridge the gap between the information that the climate scenario models currently provide and that required by the impacts research community (Wilby & Wigley 1997; Wilby *et al.* 1998a, 1998b).

Southam *et al.* (1999) used scenario analysis to examine the regional impacts of climate change on the water supply and demand issues in the Grand River basin. Grand River is the largest Canadian tributary to Lake Erie located in the southwestern region of Ontario, and is expected to be one of the more sensitive areas in Ontario to the warmer and drier conditions that may result from anthropogenic climate change. The basin has a large and growing urban population, is dependent on the Grand River for wastewater assimilation and for some municipal water supplies, has experienced significant droughts in the past century, is heavily regulated, and is located well inland from alternative sources of water such as the Great Lakes. A water use analysis model was used to downscale the data from 21 scenarios of the future surface water supplies, streamflow regulation, population and water. The ability of the system to maintain adequate streamflow (target flows) at particular sites was assessed for each scenario. Based upon the findings of the study, adaptive modifications to existing operating procedures and additional reservoir capacity were shown to provide moderately successful adjustments to all but the most severe streamflow scenarios tested.

3.2.3 Weaknesses in the Scenario Approach

There are, however, weaknesses in the scenario analysis approach when applied to climate change impacts. It can be difficult for individuals to construct scenarios that extend beyond their current beliefs, with the result being that the scenarios developed will favour the proponent's proposals while compromising other relevant factors (Schwartz 1991). Therefore, the formal process for generating scenarios that prove accurate tends to require the collaborative expertise from a diverse, multi-sectoral array of backgrounds (Kirkwood 1997; Kirkwood & Seidman 1985) such as found in the construction of the IPCC reports. While scenario construction by a diverse group of experts can be considered advantageous from many perspectives, such an approach can produce some unfortunate consequences (Kirkwood 1997). Morgan et al. (2001) demonstrated that individual experts can provide a much richer diversity of opinion than is apparent in qualitative consensus summaries such as those of the IPCC by producing estimates for climate change impacts that differed by more than 4 orders of magnitude.

One drawback to group scenario analysis can result from the apparent consensus of those who have constructed the scenarios. Models used within scenarios can be used to help make predictions, but only if they include the essential processes governing the dynamic responses (Kirkwood 1997). Although all models are necessarily approximations to reality, when so many different scenarios tend to predict similar results, this lends credence to their predictions and the general trends that they project–in spite of the inherent uncertainty. However, a bias can be introduced when the dominant modelling style misses some crucial factor(s). For example, the global circulation models are extremely unclear as to the process for appropriately modelling clouds. Similarly, forest range shifts are among the most significant potential effects of global climate change (Kempton 1991) and the consensus of forest response models to climate change is that the dieback of forests is likely on a major scale. The consequences of these predictions, and the required mitigation strategies to counteract their effects, have significant economic and policy implications. Loehle (1996) suggests that because the vast majority of forest growth models have been based upon variants of the single model of Botkin *et al.* (1972), there is the strong possibility of extremely biased predictions due to this common origin. These models might, in fact, predict forest dieback when none is likely to occur and predict range shrinkage over decades that could actually take centuries or even millennia. Because fossil fuel drives economic growth, reducing its use on the basis of similar sorts of faulty consensus projections of its destructive effects could prove economically disastrous. Hence, because climate change scenarios such as those produced by the IPCC have been based upon currently dominant beliefs, there is a non-negligible possibility that unwarranted biases have become accepted within them as the norm.

However, the main shortcoming to group scenario construction is that the scenarios produced often possess an unfortunate averaging of the anticipated system behaviour. This can be problematic for climate change studies because warning statements are often based on the notion that what matters more is behaviour of the system at its limits, not its behaviour on the average (Hammitt & Shlaykhter 1999). For instance, the risk of a shift toward calamitous extreme heat waves can double or triple under a CO_2 doubling and such outcomes may be masked in the consideration of averages (Ausebel 1991). Also, as the climate warms, there is an increased likelihood that abrupt or unexpected changes to the entire climate system could occur (Harvey 2000) and one of the critical issues in projecting future climatic change is the possible occurrence of these climate system "surprises" (Miller 2002). Examples of such surprises could include a shutdown of the North Atlantic Gulf Stream or a dramatic alteration to the frequency and intensity patterns of the El Nino-Southern Oscillation (ENSO). The Gulf Stream is responsible for transporting large amounts of heat to Western Europe and the ENSO phenomenon has a major influence over North American weather patterns, including a link to drought cycles. Other potential sources of surprises include a breakdown of the vertical salinity structure in the Arctic Ocean (leading to an abrupt disappearance of any remaining pack ice), an abrupt increase in climate sensitivity, abrupt changes in precipitation patterns or an unexpected release of methane from methane clathrates. There is evidence that some of these events have occurred in the past, but their likelihood of recurrence is difficult to estimate (Miller 2002). While the likelihood of these events is very uncertain, there would be severe consequences should they actually occur. The IPCC scenarios run the risk of not containing some of the non-negligible calamitous extreme eventualities that individual experts might have identified. Thus, contingency plans must be developed within major projects to account for high-consequence low-probability events that might not have been reflected in group-generated scenarios. However, since experience and scientific research have continuously shown that decision-makers consistently demonstrate a strong tendency to underestimate the uncertainty about the future (Dawes 1988; Kirkwood 1997), a well-constructed scenario analysis can provide a significantly better understanding of the potential consequences arising from the uncertainty and can significantly improve the alternatives designed to address the uncertainties that might exist.

3.3 Sensitivity Analysis

Both direct and indirect approaches have been broadly utilized for assessing the potential effects and interactions resulting from climatic changes. The direct approach commences from some specific climatic change scenario and examines the associated impacts that cascade down from it. Any conclusions drawn from this direct approach have inherently depended upon the model validity of the specific scenario employed in the analysis and the outputs generated by most GCM models have not proved conducive to this style of impact analysis (Harvey 2000).

The alternative, indirect approach identifies *sensitivities* or areas of vulnerability in the studied system of interest and progresses up to the climatic changes which impact these sensitivities. Sensitivity analyses focus on two related styles of what-if questions for the impacts or concerns within a given study: (1) What change in parameter *x* would cause a certain level of impact, and (2) if parameter *x* changed by an amount *y*, what would be the resulting effect? An example of the first type of question might be to determine what amount of change in precipitation would cause either a 10% increase in downstream floods or the collapse of the downstream fishery. This style of questioning emphasizes the identification of thresholds which, when crossed, can lead to significant system impacts and the specific goal is to determine where the threshold vulnerabilities occur within the system, rather than on prediction. One important climatic threshold related to agriculture concerns the vulnerability of rice crops, for which yields decrease by a factor of 3 if the mean daily temperatures increase from 25 C to 31 C at any point during the growing season (Smit & Yunlong 1996; Harvey 2000). An example of the second style of question might be to determine what the net effect on streamflow, power production, downstream flooding and fisheries would be if precipitation were to increase or decrease by 5%, 10%, or 20%? By studying the effects of various parameters changes, it may be possible to separate those specific elements that most affect the system (and should therefore receive greater scrutiny) from those elements that have only minimal impact.

The strength of sensitivity analysis is that it produces values that are independent of the correctness of any single climatic change scenario and also serves to identify the most important climatic variables which need to be provided as output from GCMs for use in the more direct-styles of analytical approach. Hence, *sensitivity analysis* is an analytic procedure which quantifies the impact of variations in model input data and parameters on model outputs (Campbell *et al.* 2001). Sensitivity analysis provides insight into the potential influence of all types of changes in inputs and assists in discriminating across parameters according to importance for the accuracy of the outcome. In its simplest form, it studies the effect on one or more output measures due to a change in only one input variable at a time, while holding all other inputs constant. However, the approach can be easily extended to include the simultaneous variation of multiple inputs. This sensitivity process identifies the important output variables, and then determines which inputs have the greatest impact upon them. Careful attention is paid to the computation of ranges over which results are considered to remain valid. The essence of this approach captures the basic what-if questioning style frequently associated with both spreadsheet-style analyses and many other forms of mathematical modelling.

Sensitivities can also be derived using composite measures to indicate different thresholds of risk level, such as the risk of subsidence due to permafrost thawing. Climate change scenarios have consistently predicted a marked warming at high latitudes (Harvey 2000, Kattenburg 1995) and Nelson *et al.* (2001) described the hazard potential associated with thawing permafrost under conditions of global warming in the Northern Hemisphere indicating that the vulnerability to subsidence is widespread. A sensitivity index was used to classify contemporary permafrost according to its vulnerability to thaw subsidence under global warming and areas at risk were partitioned into those with low, moderate or high susceptibility thresholds according to this sensitivity measure. Much of the existing infrastructure in northern regions was shown to be located in areas of high hazard potential and a geographic zone of high-risk was shown to extend discontinuously around the Arctic Ocean. Within this region were, for example: population centres in North America (Barrow, Inuvik); river terminals on the Arctic coast of Russia (Salekhard, Igarka, Dudinka, Tiksi); high-hazard transportation and pipeline corridors in northwestern North America; major natural gas production complexes and their associated infrastructure in northwest Siberia; several large population centres, extensive road networks, and mainline railways in Siberia and Eastern Russia; and the Bilibino nuclear power station in the Russian far east (Nelson *et al.* 2001). The role played by sensitivity analysis in this study was to identify the different thresholds of temperature change (inputs) that would lead to changed risk classifications (outputs) within the studied regions.

Sensitivity analysis can be used to augment confidence in models and predictions by providing an understanding of how the model response variables respond to changes in the inputs, model structures including the data used to calibrate them, and numerous other factors. Sensitivity analysis enhances model analysis in two key ways: It helps to identify the critical variables that impact the outputs of interest and is generally a straightforward process to perform on a modern computer. Since all estimates contain some errors, sensitivity analysis can be used to help understand the relative importance of each input variable and which inputs to concentrate on in further studies.

The ability to manipulate sensitivities can complicate any analysis, since the facility to generate numbers far exceeds the capacity to process them effectively. While sensitivity analysis quantifies assessments of the most important variables, it essentially necessitates a "brute force" approach in which the input variables are changed one at a time. The sheer volume of numbers produced by the sensitivities creates drawbacks, since it is difficult to derive meaningful information from large sensitivity matrices. The number of sensitivities that could be considered during an analysis can run into the millions in relatively simple models. Given the hundreds of input variables that exist in complicated climate models, presenting these results in sensitivity tables could result in an exponential increase to the size of the report output (Campbell *et al.* 2001). Most people can effectively assimilate on the order of 10-15 cases, after which point relationships become confounded and interfere with the decision-making process (Campbell *et al.* 2001). Obviously, the main disadvantage to sensitivity analysis is that it can potentially produce more information than could be effectively and efficiently processed. Large volumes of numbers can appear daunting and thorough explanations must accompany the tabulated results. Therefore, knowledgeable professionals need to bracket the ranges of uncertain numbers in their specialty area and the exercise of sound professional judgment, particularly within an effective team approach, can reduce the number of computer runs needed for decision purposes. The net result should be an organized, logical approach toward uncertainty estimation in contrast to a potential "calculational orgy" that some might purport to represent risk analysis (Campbell *et al.* 2001).

While non-linear systems can often be characterized by complex interrelationships between parameters, sensitivity analysis does not encourage analysts to consider dependencies between parameters and the likelihood that certain values will occur together. In a climate system, factors such as cloud cover, evaporation, precipitation, and temperature work both independently and in conjunction to affect various natural phenomena such as streamflow. This dependence between components is significant and must be recognized in order to generate the consideration of meaningful sensitivities. However, recognizing the dependencies between variables, substantially increases the required number of sensitivities. Skilled design of sensitivity analyses and strong knowledge of climate systems may assist in partially surmounting these shortcomings. Dependencies and restrictions can be accounted for by using creativity in structuring, synthesizing, and communicating the information that is obtained in the volume of numbers produced by the sensitivity analysis. However, while traditional sensitivity analysis can be of value, its utility can be limited by the absence of meaningful direction in how to use any resulting insights to communicate, persuade or drive change (Grossman, 2001) and no optimal sensitivity analysis procedure has been shown to exist (Clemen & Reilly 2001).

3.4 Probabilistic Analysis

An effective quantitative risk assessment should capture much more than the ranges of realistic possible outcomes; it also needs to characterize the nature of how identified risks "behave" between the identified extremes. Traditional sensitivity analysis generates range and threshold information on the consequences of changes in variables, but ignores the likelihood of these changes (Campbell *et al.* 2001). When thousands of changes for each of hundreds of variables (considered either one at a time or in combination) are possible, the number of sensitivities can clearly increase beyond the processing time available or the capacity to convey the outcomes in any meaningful way. The process of sensitivity analysis neither acknowledges the extent of uncertainty in the inputs nor does it require an assessment of how likely it is that specific values of the parameters will actually occur. Because these inherent uncertainties may permeate throughout an analysis, it can prove extremely challenging to determine the important drivers behind individual uncertainties, and users tend to remain oblivious to this analytical trait unless it has been explicitly communicated to them. Therefore, uncertainty makes sensitivity analysis a dubious technique to apply in isolation of probabilistic methods and, unless the data have been expressed in combination with a probability analysis, the chance of specific events occurring remains unclear to the user. To counter such difficulties, sensitivity analysis in climate change has evolved to incorporate conceptual models of uncertainty and the Monte Carlo simulation process (also called "stochastic simulation") has been employed to fill this void.

Monte Carlo simulation (or "simulation") is a technique for generating estimated probability distributions of outputs when the probability distributions of the stochastic inputs are either known or have been approximated (Campbell *et al.* 2001). Simulation models incorporate complex variable interactions along with the variability within the data in order to generate a random sample of the potential outcomes of a model (Grey 1995; Campbell *et al.* 2001). However, by specifying each data input to vary in a controlled and prescribed manner, realistic ranges and likelihoods of possible outcomes can be subsequently documented. Rather than producing just a single deterministic result, a much fuller spectrum of possible outcomes emerges. Monte Carlo simulation generally yields more complex descriptions of alternatives than both scenario and sensitivity analysis, since the outputs can be expressed in a distributional format (Murtha 1995). Monte Carlo simulation "answers" generally do not include a specific decision, but rather present ranges of possible outcomes and their likelihoods for various intermediate and bottom-line variables. The principle decisions associated with simulation are then left to those who examine the uncertainties represented by these simulation outputs. This distribution of variability in measures of output values encapsulates most peoples' perception of risk (Campbell *et al.* 2001) and simulation has become the preferred research tool-of-choice for analyzing many complex stochastic contexts (Law & Kelton 1991; Jones 1991).

Monte Carlo simulation provides a technique with a long history and a wide range of applications (Law & Kelton 1991; Kleijnen & van Groenendal 1992). The simulation method consists of four basic steps: (i) the identification of the risk factors; (ii) the appraisal of the likely range and probability distributions of these risk factors; (iii) the simulation of system performance with parameters sampled from the probability distributions developed for the various risk factors; and (iv) the summary of results of the analysis in a risk profile for the system performance measures or criteria (Jones 1991). The front end of a simulation model contains stochastic descriptions of the uncertain values of each of the detailed input characteristics. At the heart of the process is the model logic linking detailed system interaction characteristics to the overall outcomes and outputs of concern. The model generates a sample of the possible outcomes in a fashion that directly reflects their likelihood of occurring. While the specific mechanics of repeatedly generating random values for the inputs, evaluating the model, storing the results and producing a summary of the output, can be implemented in several ways, the core structure always remains the same (Grey 1995).

Monte Carlo methods have been extensively employed to assess the potential impacts on different systems from the likelihoods of climate change events. A relatively small increase in average planetary temperatures can be accompanied by large increases to the number of days exceeding certain thresholds (Harvey 2000; Hennessy & Pittock 1995; Wigley 1988) and climatic warming will be most perceptible as a change in frequency of these climatic extremes (Harvey 2000). While many systems may prove quite resilient to any increase in average temperature induced by climate change, they may fail under the stress of more frequent higher temperature perturbations. Variability has received increased attention in climate change research (Colombo *et al.* 1999; Katz & Brown 1992; Smit & Yunlong 1996; Harvey 2000; Hennessy & Pittock 1995; Wigley 1988) and, even when variability remains unaltered, an increased average temperature can result in the increased frequency of extreme conditions (Colombo *et al.* 1999). Even small changes to mean annual temperatures can produce dramatic changes in local climates and all of the related climate dependent activities.

Power utilities must explicitly incorporate climate change into their strategic planning processes, since the stress placed on energy supply systems during summer months is expected to be greater during extreme temperature events. Colombo *et al.* (1999) combined sensitivity analysis with Monte Carlo simulation to study the impact of climatic warming on daily peak power demand for summer extreme event frequency at nine sites across Canada by examining concurrent sensitivities to both average temperature and the daily variability. Their results indicated that while average peak power demand would not move drastically, the number of high energy consumption days would increase appreciably due to higher variability and such an outcome would place considerable stress on the provincial power utilities to satisfy this higher level of demand. The demonstrated sensitivity to both the mean and spread of the simulated temperatures underscored the risks of a greater frequency of high power demand days despite only modest changes in the average summer daily peak power demand. The simulations in their study served a descriptive purpose by demonstrating that design considerations for power supply infrastructure (and many other engineered systems) will need to include climate change concerns due to its significant potential impact on consumption (Colombo *et al.* 1999).

The results from this simulation study can also prove useful in the strategic planning process of provincial utilities. Since the daily demand for electricity continuously changes with both a diurnal and a seasonal cycle, power utilities must be prepared to continuously and instantaneously vary their supply in order to meet fluctuating demand. The electricity supply can be broken into a *baseload* supply which is sufficient to meet the minimum demand (and, hence, is always running), and a *peaking* supply which comes on-line as needed. The baseload supply in Ontario is mostly nuclear and hydro, while the peaking supply is provided by coal, gas, oil, and intermittent hydro. One implication of this is that it costs more to meet peak demand than baseload demand, because peaking capacity is used only part of the time and because its efficiency is often lower than the baseload supply (this is one reason why Ontario Hydro is moving to time-of-use rates). A second implication is that for utilities with non-fossil fuel baseload, changes in peak demand will impact CO_2 emissions more than changes in baseload (Harvey 2000). Hence, the results from the Monte Carlo-sensitivity study of Colombo *et al.* (1999) can be directly extended to indicate that climate change will increase the summer power costs in Ontario and that this increase will be accompanied by an increase in greenhouse gas emissions.

Traditional arguments against the use of simulation have focused upon the time and computing power required to develop and run the models. With current software and hardware, existing models can be expanded to include risk just by adding information on input variability. The solution time can be as low as a few seconds up to several hours depending upon the hardware and model complexity. A second complaint has dealt with the inability to link specific combinations of inputs to each studied output. Since each output may result from different combinations, answering such questions is not always easy. However, most simulation programs now provide output in sufficient detail to address such questions. Sensitivities have also been added to the simulations to communicate the relative importance of specific variables (Campbell *et al.* 2001).

Monte Carlo simulation makes explicit what is implicit in many qualitative appraisals, by attaching numbers that, at a deep level, guide thought about the prospects of a particular situation. Inconsistencies in qualitative appraisal can be "ferreted out" in the model validation phase, and the very process of imputing ranges and probabilities to variables forces a deeper integration of information and understanding about the process in question (Jones 1991). Risk assessment is mainly concerned with the spread of values and their relative likelihoods, rather than the central tendency of an uncertain quantity. Monte Carlo simulation easily provides this wider view of uncertain quantity estimation (Grey 1995) and has proved an invaluable tool for assessing potential impacts from climate change.

3.5 Other Methods: Fuzzy Logic

Fuzzy logic methods are gaining acceptance as an alternative mechanism to aggregate uncertain values. Their main application seems to be for situations where the average outcome is the main point of interest, such as in the control process parameters of household and industrial machinery. However, risk assessment for climate change is more concerned with addressing the range of values and their relative likelihoods, rather than the central tendency of an uncertain quantity. Since Monte Carlo simulation provides this wider perspective of uncertainty and it is more straightforward to apply analytically (Grey 1995), there appears to be little incentive to apply fuzzy logic in the study of climate change.

3.6 Incorporating Qualitative Factors

The methods described above are generally used when there are quantitative inputs and outputs. However, there are many impacts that can be assessed only in qualitative terms, often because of the lack of information needed to explicitly quantify them. For example, the impact of a hydroelectric project on downstream recreational fishing may only be predicted in terms of its general direction (increase or decrease) and significance (low, moderate, or high). For the hydroelectric project example in the next chapter, such qualitative issues are examined together with possible ways for addressing these qualitative factors within scenario, sensitivity and probabilistic analyses.

3.7 Conclusion

Depending upon the factors of interest within a particular application setting, each of the three basic methods described above (or combinations thereof) can prove more advantageous in addressing uncertainties than the other methods. For a project impact being examined under an environmental assessment, these factors of interest could include such things as: (i) the type of impact of the project; (ii) the contribution/reduction of the project to GHG emissions; (iii) the effects of climate change on the project; (iv) the effects of climate change on the impacts of the project; (v) whether an impact is well defined and quantitatively measurable or ill-defined and qualitatively measurable; (vi) whether the likelihood of the impact can be probabilistically defined or is unknown; (vii) the level of understanding about the relationship between climate change and the impact; (viii) the difficulty required in the use the method, e.g. expertise, data and cost, and; (ix) the importance of the information resulting from the analysis. For example, since GHG emissions can be quantitatively measured and uncertainties about emissions from a manufacturing process can be the result of uncertain factors of production, sensitivity analysis might be the most appropriate approach to take for analyzing the impacts. Conversely, to realistically estimate the GHG emissions from vehicular travel in an urban area, might require the development of alternative urban development scenarios combined with a subsequent sensitivity analysis on the vehicle emissions arising within each scenario.

The hydroelectric example to be introduced in the subsequent chapter illustrates the use of a subset of the types of impacts commonly examined in an environmental assessment and examine which of the methods prove be most appropriate to address the effects of climate change uncertainties. As is shown in this hydroelectric example, a quantitative scenario analysis coupled with a probabilistic analysis provides an appropriate approach for understanding the effects of climate change uncertainties on streamflow and energy production, since the scenarios produce quantitative estimates of the climate change variables that can be incorporated into existing stochastic streamflow and energy models. However, sensitivity analysis proves more appropriate for addressing the uncertainties about the potential impacts on downstream fisheries, since the link between streamflow and the fishery is more complex and less well understood. These effects and other related issues are examined subsequently.

4

Incorporating the Impacts of Climate Change into an Environmental Assessment: A Hydro-Electric Example

4.1 Introduction

In practice, most projects have been designed on the basis of established historical standards, but such historical specifications will most likely not remain applicable under conditions introduced by a changing climate. Consequently, since there are many ways in which climate change impacts could be adapted to at the project level, these approaches need to be readily considered during the EA. The major difficulty for project EAs is to determine how changing climatic conditions might affect the project and how to effectively incorporate any resulting uncertainties into the decision-making process. These uncertainties include impacts not only on the regional climate variables, themselves, but also on socio-economic systems. Several studies have indicated that an adequate accounting for plans and methods to adapt to the impacts from climate change has not occurred within the structure of most EAs (Murphy 2003). As described earlier, the inadequacy of current EA practices is significant, since the impacts from climate change could have a considerable effect on the relative importance of any decision criteria employed and could, therefore, affect the selection of the preferred alternative.

In general, the most needed climate information required for project EAs relates to the specific climate details for the region in which the project is being undertaken; yet, there are high degrees of scientific uncertainty associated with such geographically-specific estimates. Complicating the impact assessments required for an EA even further, there are significant uncertainties about the effect of climate change both with and without any activities introduced by the project. Thus, many proponents tend to view the scientific basis surrounding the predictions of future climate to be too uncertain to have sufficient confidence to act upon (Lee 2001). It is therefore understandable why many proponents have either ignored the effects of climate change in EAs or have claimed that any effects resulting from their project would pale in significance when considered on regional, national or global scales. In spite of such views, there is a need, including a legal obligation, for proponents to address the potential impacts and uncertainties from climate change in their EAs. This chapter illustrates the use within an EA of the methods described in the previous chapter to address the various uncertainties arising from climate change and how the results can be presented to decision-makers and stakeholders.

4.2 Background for the Hydroelectric Example

The illustrative, hypothetical example used here is based upon an actual proposal for a run-of-the-river hydroelectric project in Northern Ontario (see Appendix 1). The run-of-the-river feature of this facility implies that the information deemed most relevant to its future energy production capabilities would be the future streamflow of the river, which, in turn, depends upon such weather variables as precipitation, temperature, wind, etc. If all of these future weather variables were "known", then well-established models could be used to calculate both the corresponding future streamflows and the facility's energy production generated from these streamflows. However, the uncertain impacts from climate change directly create uncertainties in the patterns of future weather variables, thereby causing uncertainties in the future streamflow conditions, which necessarily creates subsequent uncertainties regarding the future energy production.

In addition, these uncertainties could initiate a rebalancing of the relative priorities between various decision criteria for the facility design. For example, the hydroelectric plant might have initially been designed on the basis of power production using historic streamflow data. However, if it could be determined that changing climatic conditions might bring about increased likelihoods of extreme streamflow events, then greater design emphasis might need to be devoted to such criteria as future flood control and the future of downstream fisheries. While project design changes might have no effect on preventing the future streamflow uncertainties, it might be possible to modify the design criteria to reduce the effects and vulnerabilities of the remaining uncertainties. Hence, the extent of such possible impacts and criteria should be sufficiently addressed during the project's EA.

In addition to the effects of climate change *on* the project, certain stakeholders may wish to have the project's implications *for* climate change addressed in the EA–particularly its contributions to GHG emissions. While acknowledging that even infinitesimal emissions may contribute to climate change, proponents might view GHG impacts on climate as a larger issue more appropriately addressed through government policies and programs. However, broad national emission reduction targets to achieve Canada's requirements under the Kyoto protocol have been established for various industries (Canada 2002). Specifically, of 65 megatonnes of CO_2-equivalent emissions reduction required by Canada's "large industrial emitters", the apportioned target (16%) for the electricity generation sector is 13 megatonnes (Canada 2002). Hence, to satisfy the technical disclosure issues of both proponents and stakeholders in the EA, the hydro project's anticipated GHG contributions could be disclosed relative to this industry sector target as a proxy measure for the climate change impact of the project.

A key finding from the CICS review described in Chapter 1 included the recommendation that a range of probable futures resulting from climate change needs to be made available to support sensitivity studies in relevant project EAs and that there is a need to present information on the reliability/ uncertainty of these projections (Murphy 2003). For the hydroelectric example, these probable climate change futures will consist of historical streamflows modified by the methods of (i) scenario analysis, (ii) probabilistic analysis, and (iii) sensitivity analysis. To evaluate the reliability/uncertainty under each of these potential futures in the example, the climate change impacts *on* the project will examine both streamflows and energy production, while the climate change impacts *of* the project will address greenhouse gas emissions, flooding, and fisheries.

4.3 Baseline Streamflow and Energy Production Using Historical Data

Prior to exploring possible future climate impacts using each of the methods mentioned, it proves beneficial to consider the project operating under some "baseline" condition for comparative purposes. Since the study upon which this example was based contained information on monthly rates of historical streamflows, a hypothetical baseline condition of no climate change was established via a projection of the historical data over the period 2010-2099 (the reason for this time period is explained subsequently). Viewed from an alternate perspective, this baseline condition consists of three repetitions of 30 years of prior historical data under the assumption that these historical conditions would remain in effect over the specified time horizon. In addition, the criteria for the designed production capacity of the hydro facility was based solely on the premise of a continuation of these historical conditions. Hence, a model for the baseline situation would generate the energy production of a hydro facility designed for and operating under current conditions. A schematic diagram of the modelling used to generate the resulting streamflow and energy production of this baseline case appears below.

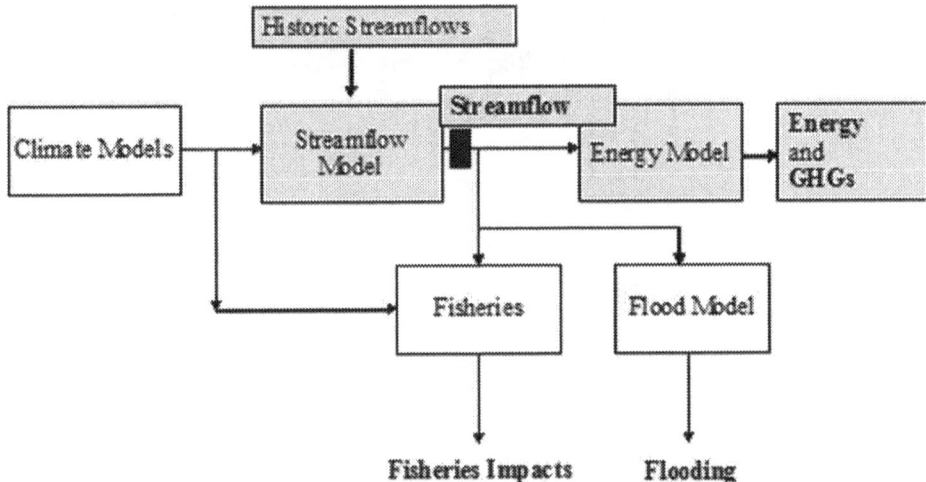

Figure 4.1: Schematic Diagram of Hydro Example based upon Historic Data

In this diagram, the shaded cells indicate the relevant components needed to generate the baseline streamflow and energy values. In this process, the historical streamflow rates are used as inputs into a streamflow model in order to create the actual streamflow values projected over the period 2010-2099. These projected streamflow values subsequently become the input of an energy model that generates the values of the energy produced by the hydro facility based upon its design criteria. Using this series of models to create the baseline condition (see Appendix 1), the average rate of monthly streamflow and the average monthly energy production for each year of the time period appear, respectively, in the following two figures.

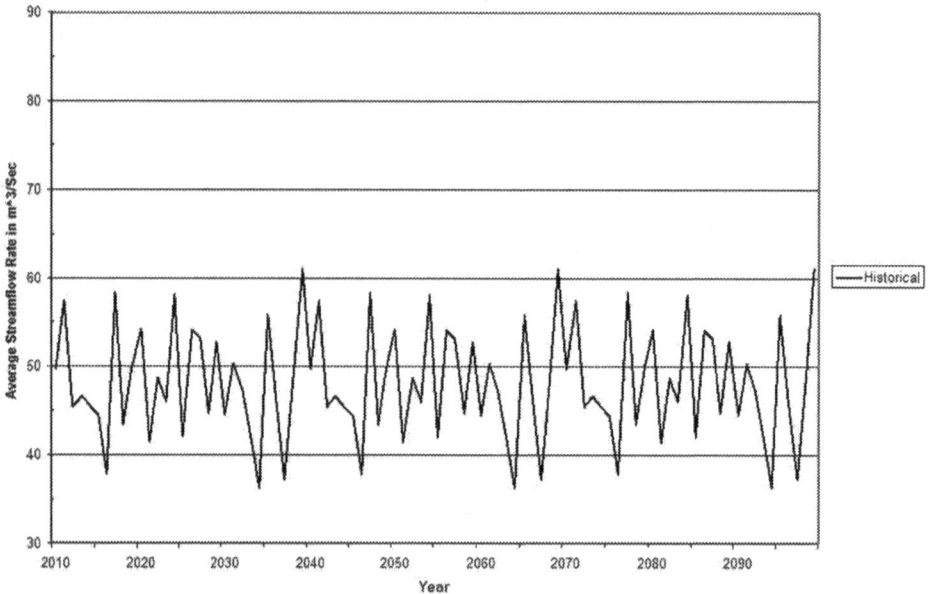

Figure 4.2: Average Rate of Monthly Streamflow under a Projection of Historical Data

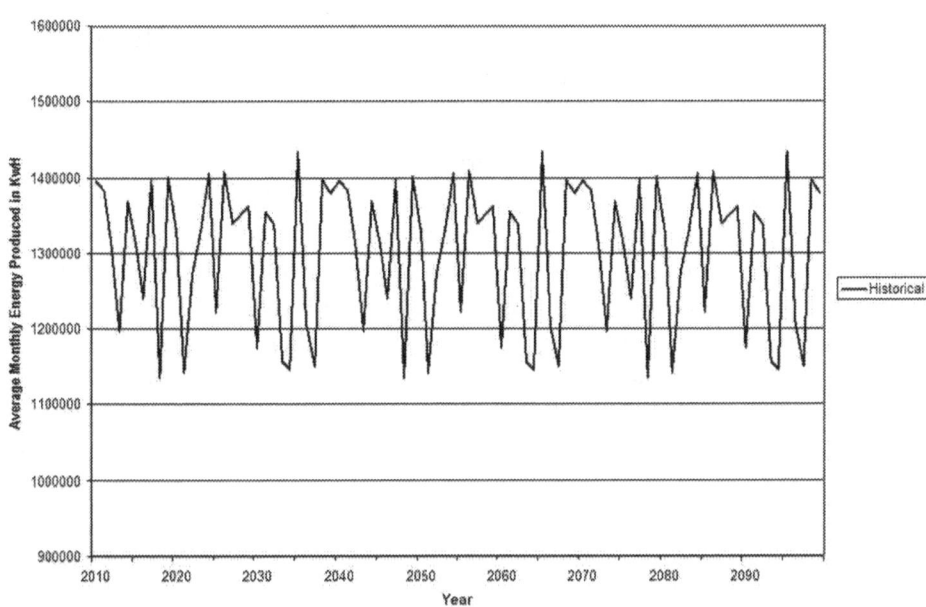

Figure 4.3: Average Monthly Energy Produced Each Year under the Historical Scenario by the Hydro Facility Designed Using Historical Data

From these projections and with the design capacity based upon historical data, the expected annual energy produced by this hydro facility would be 15,600 Mwh. Since the "cleaner" energy produced by the hydro facility could be used as a substitute for an equivalent quantity of energy generated at a fossil-fuel operated plant, the project would contribute to an overall net *reduction* in GHG emissions. Hence, in comparison to an equivalent capacity fossil-fueled facility, the expected annual reduction in GHG emissions would be approximately 14,000 tonnes of CO_2. This reduction converts to approximately 0.11% of the total electricity industry's emission target under Kyoto and this value could be used as the surrogate measure for describing the project's impact on climate change.

4.4 Streamflow and Energy Determination based upon Climate Change Scenarios

If climate change is to be incorporated into an analysis, then appropriate means and mechanisms become necessary to accomplish the task. For the hydro project, this requires the addition of "climate models" to the streamflow and energy models already employed in order to modify the projected future values of the streamflow. These climate models generate the specific values of the future weather variables which now become the inputs into the streamflow model (replacing the historic streamflow values used in the prior baseline case). The shaded cells of the modified schematic diagram below indicate the relevant components now used to generate the requisite streamflow and energy values.

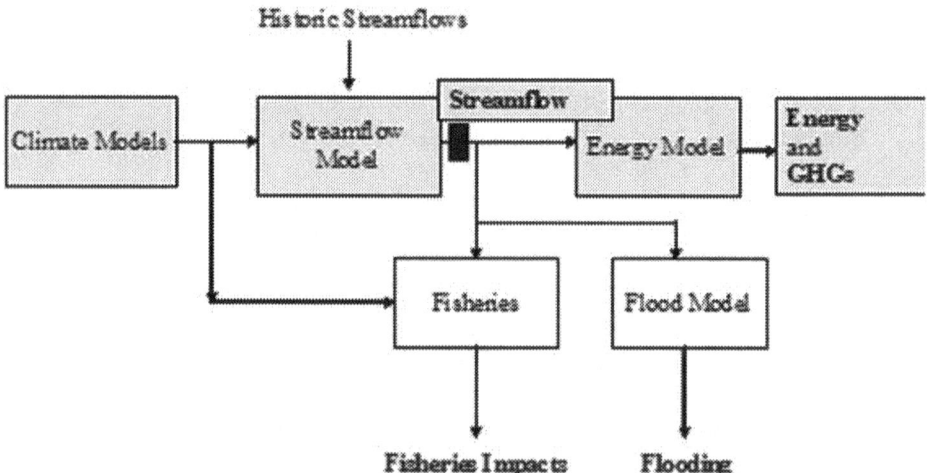

Figure 4.4: Schematic Diagram of Hydro Example based upon Climate Change Data

As indicated in the review-of-methods chapter, the predominant methodology for modelling climate change is scenario analysis. Hence, in this section, scenarios will be used for the "climate models" component in the above diagram.

There are numerous scientific institutes throughout the world that have conducted extensive scenario analyses of the future climate. To access the results of these studies, the website for the Canadian Climate Impacts Scenarios (CCIS 2003) permits users to download climate scenario data produced both from their own climate models and also from many of the other major international climate institutes. The computer screenshot below shows a gridded map of North America appearing on the CCIS website that provides one entry point for accessing the available data. In order to download climate data, the user must select the specific grid corresponding to the region from which the data is desired.

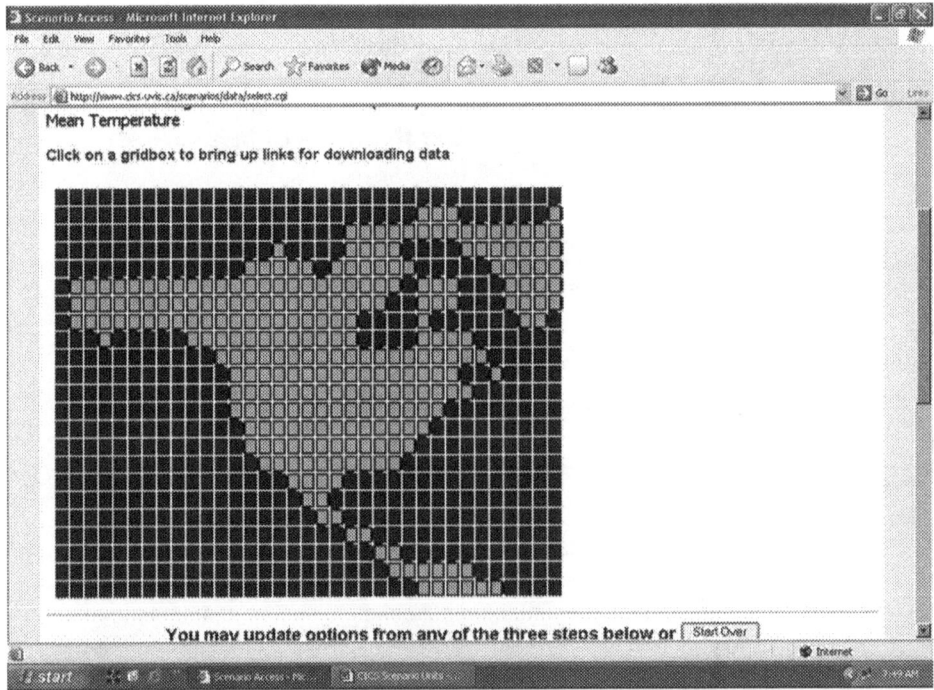

Figure 4.5: Computer Screenshot of Gridded Climate Map on the CCIS Website

The data from each scenario is partitioned into several different time slices of 30 years each: 1961 to 1990; 2010 to 2039; 2040 to 2069; and, 2070 to 2099. The 1961-1990 time slice corresponds to historical weather data collected from the specific gridpoint, while the remaining three times slices provide the future weather values generated by the scenario for the referenced grid location. Hence, future weather variables can be obtained for specific geographic regions in the country over the time period from 2010 to 2099. This data availability explains why the 2010-2099 time period was employed for the historic baseline projection of the hydro project.

Weather data collected for up to seventeen different variables are available from each scenario. The possible regional variables that can be obtained, together with their units of measure, are indicated in the subsequent table. It should be noted that, other than for precipitation, cloud cover, and wind speed, the units of measure for the future weather variables are the same as those from the historical time slice. The precipitation, cloud cover, and wind speed variables need to be expressed as percentage changes to the actual values in the historic time slice in order to prevent the future values from ever taking on infeasible, negative values.

Table 4.1: **Weather Variables and the Units of Measure for Scenarios from CCIS**

Units
Note that those variables expressed as Percent of 1961-1990 have different units for the baseline than the scenario data, so that the absolute changes in future scenarios may be determined.
Tables of variables and time slices available by experiment are also available.

VARIABLES	Units for 1961-1990 Scenario Data Files and all Time Series Data Files	Units for Scenario Data Files (NOT Time Series)
Temperature	Degrees Celsius	Degrees Celsius
Precipitation	mm/day	Percent of 1961-1990
Maximum Temperature	Degrees Celsius	Degrees Celsius
Minimum Temperature	Degrees Celsius	Degrees Celsius
Specific Humidity	kg/kg	kg/kg
Incident Solar Radiation	W/m^2	W/m^2
Cloud Cover	Fraction	Percent of 1961-1990
Wind Speed	m/s	Percent of 1961-1990
Evaporation	mm/day	mm/day
Soil Moisture	capacity fraction	capacity fraction
Mean Sea Level Pressure	hPa	hPa
Snow Water Content	kg/m^2	kg/m^2
Sea Ice	kg/m^2	kg/m^2
Vapour Pressure	hPa	hPa
Relative Humidity	Percent	Percent
Diurnal Temperature Range	Degrees Celsius	Degrees Celsius
Surface Temperature	Degrees Celsius	Degrees Celsius

Unfortunately, several of the scenarios on the CCIS website either contain restricted data sets (i.e. do not include all seventeen possible variables) or do not possess variable estimates for all of the future time slices. Since the user would necessarily be restricted to using only those scenarios which contained information for all of the required climate variables over the applicable time horizon, the first step in scenario selection is to identify exactly which scenarios meet the desired requirements for the needed variables. However, this vital information on the data availability of each experiment operating under each specified scenario by each climate institute (the "model") is provided by the CCIS in the form of the following two tables.

Table 4.2: Presence or Absence of Particular Weather Variables in Specific Scenarios

Legend: x - variable is present

Model	Emiss. Scen.	Variables										
		temp	prec	tmax	tmin	shum	rad	tcld	wind	evap	soil	mslp
CGCM2	SRES	x	x	x	x	x		x	x	x	x	x
GFDLR30	SRES	x				x						
ECHAM4	SRES	x				x		x			x	
CSIROMk2	SRES	x	x	x		x		x		x	x	
CCSRNIES	SRES	x	x	x		x		x			x	
HadCM3	SRES	x	x	x		x		x			x	x
NCARPCM	SRES	x	x	x		x					x	
CGCM1	IS92a	x	x	x	x	x	x	x	x	x	x	x
CGCM2	IS92a	x	x	x	x	x		x	x	x	x	x
HadCM2	IS92a	x	x	x			x	x			x	
GFDLR15	IS92a	x				X					x	
ECHAM4	IS92a	x	x	x		X		x			x	
CSIROMk2	IS92a	x	x	x		x					x	
CCSRNIES	IS92a	x	x	x	x			x			x	
HadCM3	IS92a	x				x		x			x	
NCARPCM	IS92a	x	x	x		x					x	

Table 4.3: Availability of Weather Data in the Particular Time Slices of Specific Scenarios

Model	Emissions Scenario	Experiment	Time Periods			
			1961-1990 Baseline	2010-2039	2040-2069	2070-2099
CGCM2	SRES	A21	x	x	x	x
		A22	x	x	x	x
		A23	x	x	x	x
		A2x	x	x	x	x
		B21	x	x	x	x
		B22	x	x	x	x
		B23	x	x	x	x
		B2x	x	x	x	x
CSIROMk2b	SRES	A11	x	x	x	x
		B11	x	x	x	x
		A21	x	x	x	x
		B21	x	x	x	x
GFDLR15	IS92a	ga1	x	x		
		gg1	x	x		
ECHAM4	IS92a	ga1	x	x		
		gg1	x	x	x	x

Another complication can arise due to different scale resolutions used by the different climate institutes for constructing their scenarios. Details on the different scale resolutions of the scenarios can also be found on the CCIS website and the following table provides an example of the number of grid points used to represent Canada (and hence the resolution) in a small cross-section of different climate experiments.

Table 4.4: Scale Resolutions of Specific Scenarios

GCM	CGCM1	CGCM2	HadCM2	ECHAM4	CCSR98	CSIROMk2b	GFDL-R15
GCM type	Spectral T32	Spectral T32	Finite Grid	Spectral T42	Spectral T21	Spectral R21	Spectral R15
AGCM resolution °lat×°long	3.75×3.75	3.75×3.75	2.5×3.75	2.8×2.8	5.6×5.6	3.2×5.6	4.5×7.5
AGCM number of vertical levels	10	10	19	19	20	9	9
Global grid: number of lat×long boxes	48×96	48×96	73×96	128×64	64×32	64×54	48×40
Canadian window: number of lat×long boxes	13×35	13×35	20×35	18×47	9×24	15×24	11×19
OGCM resolution °lat×°long	1.8×1.8	1.8×1.8	2.5×3.75	2.8×2.8	2.8×2.8	3.2×5.6	4.5×3.7
OGCM number of vertical levels	29	29	20	11	17	21	12
Flux correction	Yes		Yes	Yes	Yes	Yes	Yes
Warming (°C) at CO_2 doubling	2.7		1.7	1.3	2.4	2.0	2.2

These scaling differences necessitate that users must ensure that the weather data taken from different experiments contains the specific region in which their project actually occurs. For example, the above table indicates that the CGCM1 group of scenarios (Column 2) has represented Canada as a grid of 13x35 cells (Row 6), while other climate institutes have adopted different scale resolutions. Obtaining precipitation changes for the GG1 scenario from the CGCM1 group of experiments, for instance, would entail selecting the appropriate gridpoint from the following map of Canada.

An example scenario at original resolution: CGCM1 Greenhouse Gas Only (GG1) Summer Precipitation Change (%) for the 2020s.

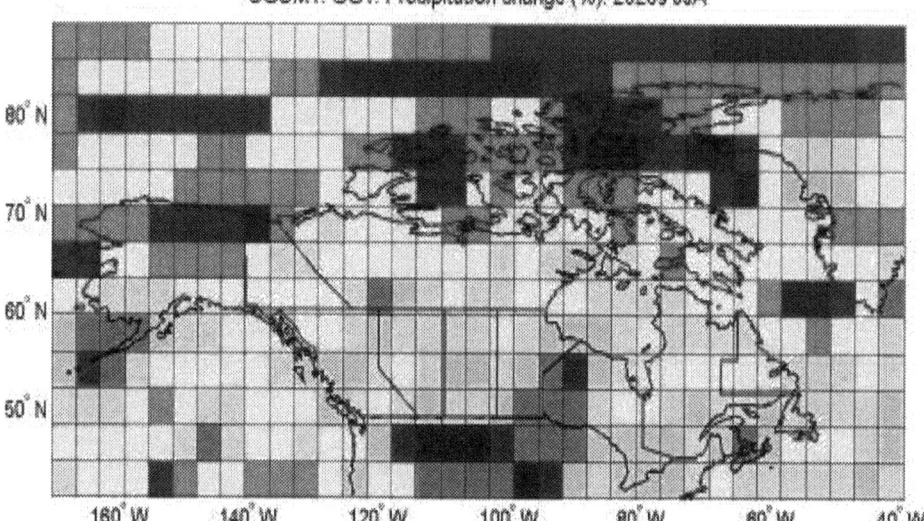

Figure 4.6: Screenshot from CCIS Website Showing Availability of Precipitation Data from the CGCM1 Scenario at the Original Resolution

Since, in general, the weather variables most needed for project EAs require very specific regional details, a number of approaches have been used to downscale the coarser-scaled scenario data from the various climate institutes into much smaller grid references (Wilby & Wigley 1997; Wilby *et al.* 1998a, 1998b). The CCIS website does include scenario data expressed at much finer resolutions than that provided by the original scenario analyses and, as an example, fine-resolution data for the CGCM1-GG1 scenario could be obtained from the following grid. However, since downscaled data has required some form of interpolation technique, users should be cautious about the precision of finer resolution data.

An example of a scenario interpolated to 0.5° latitude/longitude resolution
CGCM1 Greenhouse Gas Only (GG1) Summer Precipitation Change (%) for the
2020s

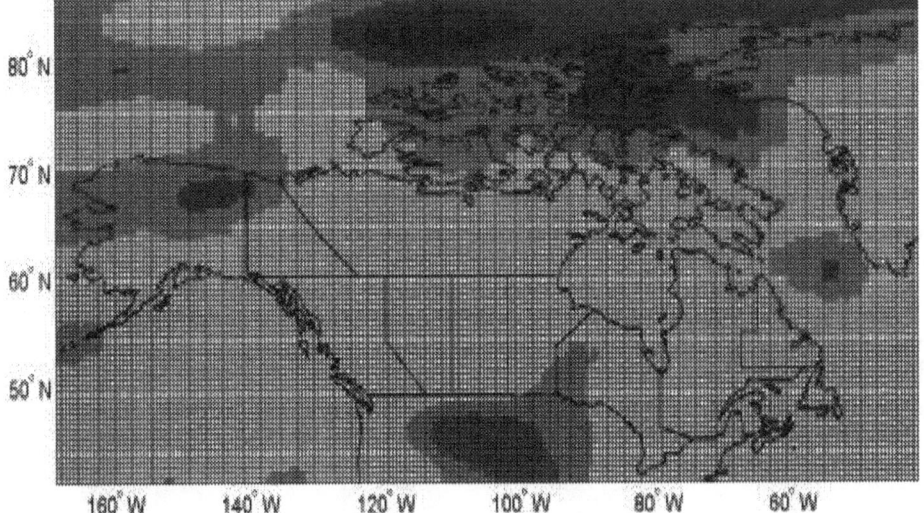

Figure 4.7: Screenshot from CCIS Website Showing Availability of
Precipitation Data from the CGCM1 Scenario at an
Interpolated Finer Resolution

Due to the resolutions used in scenario computations, it is impossible for climate models to provide certainty about specific weather variables in specific locations. Unfortunately, since highly specific weather variables are needed to determine regional impacts, any predictions of weather impacts on geographically-localized projects requires detailed climatic information in which there is the lowest level of confidence (Harvey 2000). While the CCIS provide instructions on how to employ fine resolution data, in the hydro example (and perhaps for many project EAs) the actual coarser resolution data from the original scenarios has been deemed sufficient.

Once the user has determined which scenarios possess the needed variables over the required time periods, a decision must be made as to which scenarios to include in their analysis. The IPCC suggest that scenarios be selected by a means consistent with international methodologies (IPCC 2001; SRES 2000). In any study influenced by climate change impacts, it is essential to consider a range of different scenarios so that the distribution of possible outcomes can provide a useful context for understanding the relative likelihoods of various occurrences. The recommendation from the IPCC is that "users should...apply multiple scenarios...[that] span a range of possible future climates, rather than designing and applying a single 'best guess' scenario" (IPCC 2001) and the CCIS specify that users should include scenarios constructed by at least two different climate institutes (CCIS 2003). Reviewing as many scenarios as possible provides a broader context of what is likely to happen and how key variables might change in the future.

However, should there be insufficient time to use *all* scenarios meeting the specified requirements, then specific scenarios should be selected that represent the extreme ranges of the key variables required in the analysis, as well as a more moderate, intermediate scenario (CCIS 2003). Under such circumstances, it is important to select enough scenarios to bound the full range of scenario results for the relevant climate variables. For instance, since a streamflow model depends strongly on precipitation and temperature projections, the practitioner should select four scenarios that would cover the four combinations of warm/cool and wet/dry scenarios. The CCIS website provides several useful comparative figures that illustrate which scenarios provide extreme value ranges for any two variable combination possibilities.

Using 25 scenarios which contained the future climate variables needed as inputs for the streamflow model, the following figure representing the possible future rates of the average monthly streamflow of each year (i.e. the annual average of each month's average streamflow) over the period 2010-2099 were produced for the Hydro example. It can be seen that there is considerable annual variability both between the different scenarios and within any given scenario. However, the figure contributes an visual impression of what might happen including: (i) that the range of extreme streamflows would be somewhere between 30 m^3/sec and 80 m^3/sec; (ii) that the streamflow will tend to range between 40 m^3/sec and 60 m^3/sec, in general; and, (iii) a sense that average streamflow over the entire period would be approximately 55 m^3/sec. Such knowledge provides key information concerning potential design parameters for the hydro facility and, in particular, may contribute to the establishment of vital criteria that might be needed regarding potential uncertainties that this facility might experience over the planning period. Since any of these scenarios *could* occur, it would be imperative to consider all of the design options needed to account for these uncertainties in the decision process—including due consideration of the occurrence of any "extreme" streamflow events.

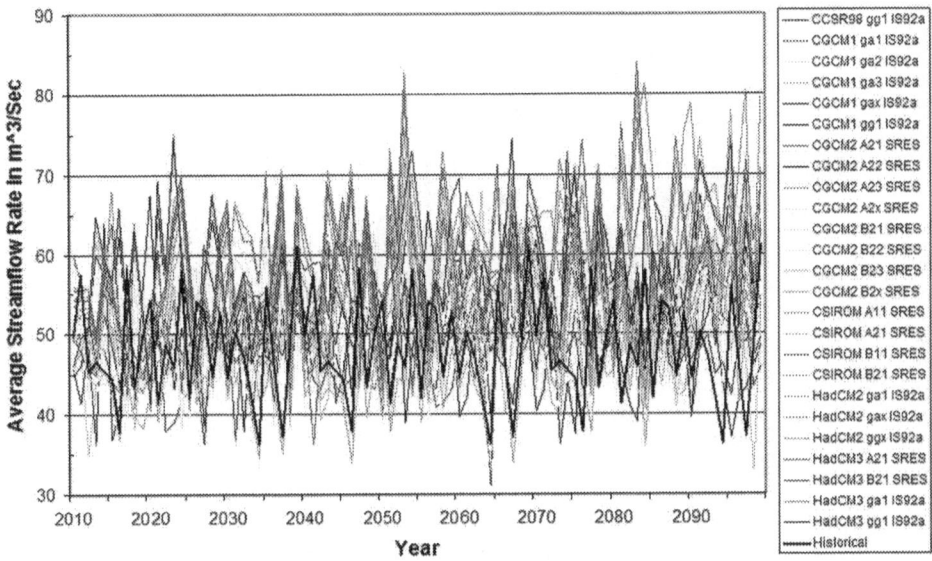

Figure 4.8: Average Rate of Monthly Streamflow under
Different Scenarios

The resulting streamflows determined from the scenarios provide the inputs required for the energy production model. However, this step proves somewhat more complex, since the design capacity for the hydro facility must be determined before these streamflows actually occur. Consequently, the design capacity for the facility becomes a key decision variable that needs to be determined for the hydro project, *a priori*, and this decision could be based on numerous different criteria–including on the basis of the possible future streamflow patterns shown above. If the capacity had been designed solely on the criteria of the known historical rates of streamflow, then the projections for the annual average monthly energy of this facility operating under the streamflow conditions of each of the scenarios would be those in Figure 4.9.

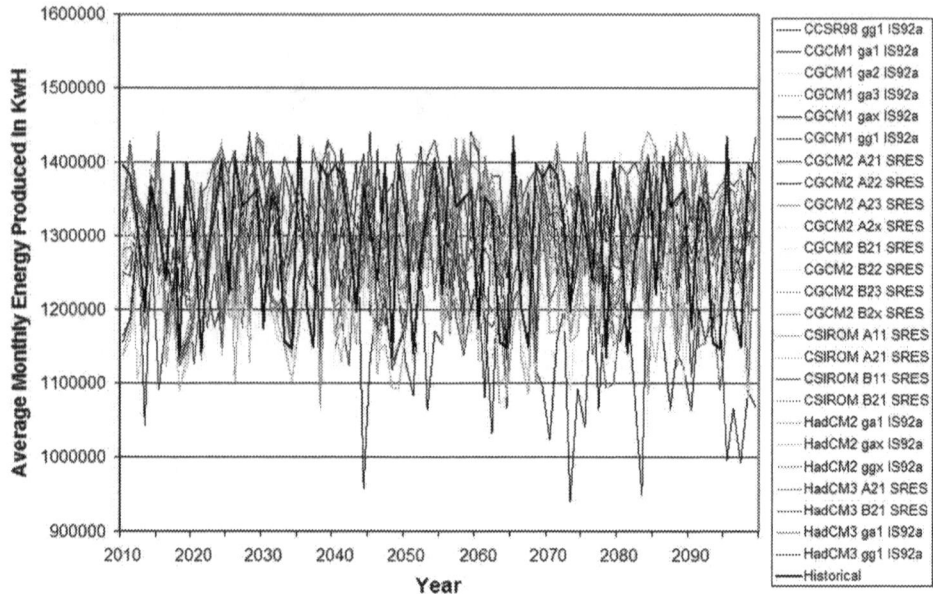

Figure 4.9: Average Monthly Energy Produced Each Year under ALL Scenarios by the Hydro Facility Designed Using Historical Data

As with the previous streamflow figure, the energy production figure provides a visual impression of the uncertainties in the possible levels of energy production over the time horizon including values for extreme ranges (particularly the minimum production), the general range of energy production, and the average energy production.

However, since the various scenarios could be considered as representations of the best projections of future weather patterns that current scientific knowledge can provide, they could reasonably be considered as a plausible representation of the range of probable futures resulting from climate change. Therefore, it makes sense to consider these possible futures in the making of such key decisions as the facility's design capacity.

While all of the scenarios employed above could (or should) be considered in such decision-making, for illustrative and visualization purposes *only* the subset of four CSIROM scenarios, together with the historical data, will be used (i.e. in the forthcoming example, the CSIROM scenarios are assumed to be representative of the complete scenario set, but actually produce results that are far narrower in breadth than had a larger, more representative set of scenarios been used). The two subsequent figures represent a repetition of the two previous figures using only this prescribed subset of scenarios.

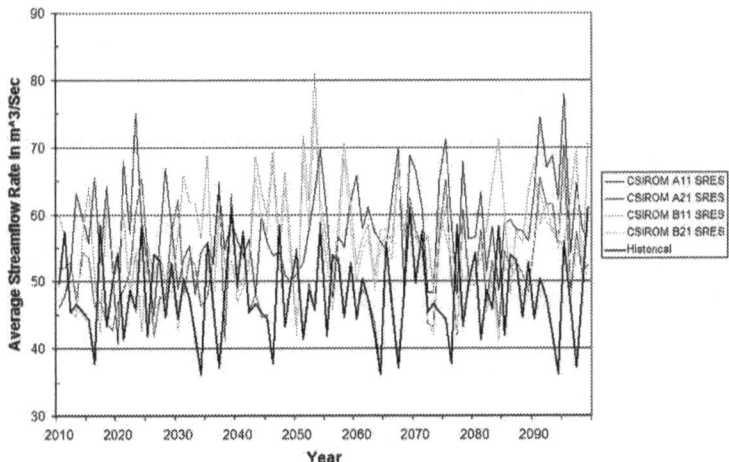

Figure 4.10: Average Rate of Monthly Streamflow under a Projection of Historical Data and Several CSIROM Scenarios

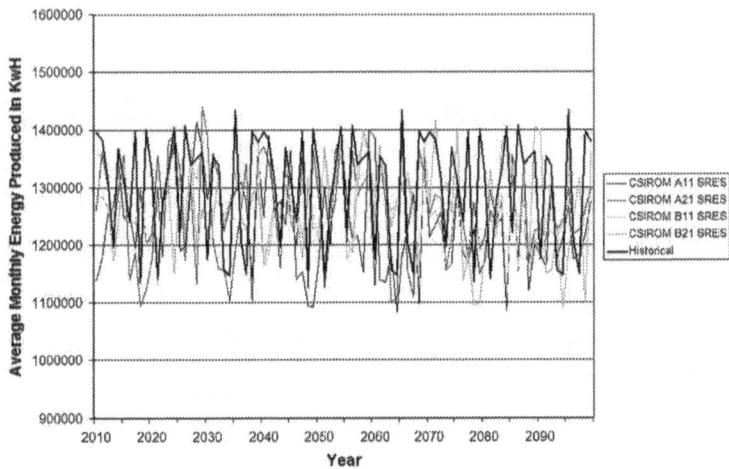

Figure 4.11: Average Rate of Monthly Energy Produced Each Year under the Historical and 4 CSIROM Scenarios by the Hydro Facility Designed Using Historical Data

Hence, if the historical data had been used to determine the design capacity of the hydro facility and the CSIROM scenarios are considered as representative of the future variability that may actually occur in the region of the project, it can be seen that the annual average monthly energy production can range anywhere from approximately 1,100,000 kwh up to 1,400,000 kwh, but could more frequently be expected to fall between 1,200,000 kwh up to 1,300,000 kwh. Since it has been assumed that no single scenario could be treated as more or less probable than others (SRES 2000), the likelihood that any one of these scenarios could occur would be the same.

To illustrate the effect of this assumption, the following table shows the mean annual average monthly energy that would have been produced by the facility over the entire time horizon (i.e. an average of the annual monthly averages) under the condition that each scenario had, in fact, actually occurred. These values provide an estimate of the uncertainty in the average performance of the facility over the time horizon. In addition, the impact on the climate for each scenario could also be determined by a comparison of the average annual reduction in GHG emissions obtained from using the hydro energy instead of fossil-fuel derived energy. Thus, it could be expected that the hydro project could reduce average annual GHG emissions by 13 kilotonnes to 14 kilotonnes. This reduction represents an average annual contribution to the Kyoto target of the Canadian electricity generation sector of between approximately 0.10% to 0.11%.

Table 4.5: Expected Monthly Energy Production & GHG Reductions 2010-2099

Scenario	Facility Capacity Based Upon Historical Data	
	Average Monthly Energy in Kwh	Annual Reduction in GHG Emissions in Tonnes CO2
Historical	1,301,307	14,054
CSIROM A11	1,239,742	13,389
CSIROM A21	1,250,407	13,504
CSIROM B11	1,261,209	13,621
CSIROM B21	1,268,823	13,703

Since all scenarios could be considered equally likely, it is of considerable interest to explore the consequences of making decisions based upon the use of any one scenario and determining the subsequent impacts should, in fact, any of the scenarios actually occur. Since the CSIROM B11 and CSIROM B21 scenarios contain, respectively, the maximum and minimum values of average monthly streamflow over the entire time horizon it is instructive to contrast the energy production resulting from having designed the capacity of the hydro facility on the basis of these "extreme" scenarios under the possible future conditions of each of the scenarios. Hence, the following figures show the energy production resulting under the conditions of each scenario had the capacity of the hydro facility actually been designed on the basis of either the CSIROM B11 or the CSIROM B21 scenario, respectively.

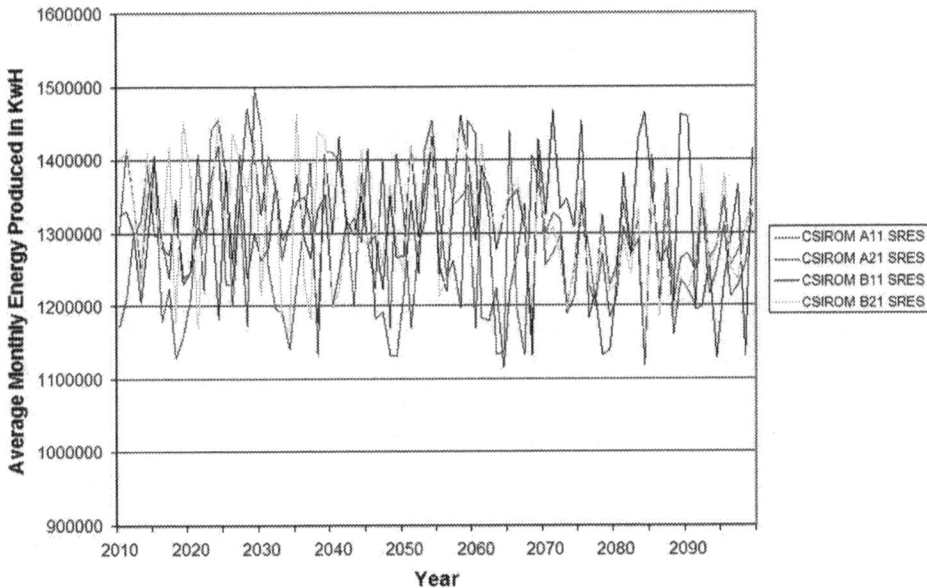

Figure 4.12: Average Monthly Energy Produced Each Year under the 4 CSIROM Scenarios by the Hydro Facility Designed Using Data from CSIROM B11 (the Scenario with the Highest Energy Output)

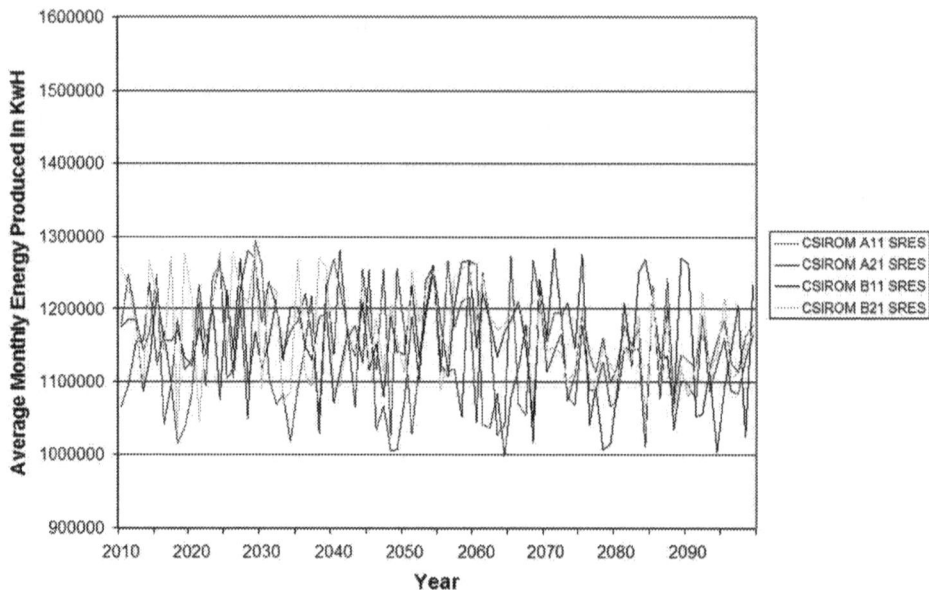

Figure 4.13: Average Monthly Energy Produced Each Year under the 4 CSIROM Scenarios by the Hydro Facility Designed Using Historical Data from CSIROM B21 (the Scenario with the Lowest Energy Output)

These graphs clearly indicate that there are considerable planning implications for energy production based upon the facility's design capacity. Various values can be calculated from the output and summaries of certain values appear in the subsequent tables. If the capacity of the facility had been designed on the basis of the "low energy" scenario, then the annual average monthly energy production would be in the 1,140,000 to 1,168,000 kwh range (i.e. approximately 1.15 Gwh). However, had the capacity been designed on the "high energy" scenario basis, the annual average monthly energy would range from 1,200,000 to 1,311,000 kwh (or approximately 1.25 Gwh). Corresponding observations can also be made with respect to the climate change impacts from the resulting average annual reductions in GHG emissions. On the basis of using the "low energy" scenario to set the design capacity, the hydro project would be able to reduce annual emissions by 12.3 to 12.6 kilotonnes, while a capacity designed on the "high energy" scenario would result in emissions reductions between 13.8 and 14.1 kilotonnes.

Hence, if the CSIROM scenarios were broadly considered to represent *all possible* climate change scenarios (an unrealistic assumption), then this analysis would be able to indicate that under all possible climate outcomes, this hydro facility would be able to produce between 1.1 and 1.3 Gwh on an annual average monthly basis. Using a similarly simplified assumption, the hydro facility would be able to reduce the annual GHG emission under all climate outcomes by between 12.3 and 14.1 kilotonnes, which corresponds to a 0.09% to 0.11% reduction to the Canadian Kyoto target for the electricity generation sector.

Table 4.6: **Expected Monthly Energy Production in Kwh for 2010–2099**

| | Facility Capacity Design Based Upon Data From: | |
| | CSIROM B21 | CSIROM B11 |
Scenario	Scenario	Scenario
Historical	1,194,325	1,344,701
CSIROM A11	1,140,190	1,282,184
CSIROM A21	1,152,737	1,291,820
CSIROM B11	1,158,754	1,305,290
CSIROM B21	1,168,313	1,311,126

Table 4.7: **Expected Annual Reduction in GHG Emissions in tonnes of CO_2**

| | Facility Capacity Design Based Upon Data From: | |
| | CSIROM B21 | CSIROM B11 |
Scenario	Scenario	Scenario
Historical	12,899	14,523
CSIROM A11	12,314	13,848
CSIROM A21	12,450	13,952
CSIROM B11	12,515	14,097
CSIROM B21	12,618	14,160

It would appear that the higher capacity facility will always tend to produce more energy than the lower capacity facility under all future scenarios. However, this higher output could perhaps be counterbalanced by economic considerations based upon the costs of operating a larger plant and on the possibilities of times during which the facility would not be operating at full capacity. Or there may be an indication that the criteria for determining design capacity requires an appropriate re-evaluation.

Irrespective of these considerations, this type of analysis demonstrates that the scenarios could be used in numerous ways to address different decisions and that the information produced provides data on the ranges, expectations, and uncertainties inherent in the operation of the facility in the face of a changing climate. If these scenarios could be considered as representative of all possible future climate change paths, or at least their extremes, then the values determined in this output would provide the extreme ranges of the energy production and GHG emission reductions possible for this hydro facility. The determination of such limits would prove to be essential information in an EA.

This section has introduced the use of scenario analysis into the planning process for the hydro project and has demonstrated that scenarios can be used to produce the representative range of climate change futures recommended by the CICS to support the analysis in EAs. The information produced can be used to support the planning implications from using various different decision criteria for designing the hydro facility (i.e. best-case-worst-case analysis, decisions based on averages or ranges of values, etc.), while simultaneously being used to answer questions on the environmental implications arising from such decisions. Furthermore, it has also been demonstrated that an abundance of quantitative and graphical data can be produced that would provide both EA proponents and stakeholders with considerable information on the possible reliabilities and uncertainties that could be encountered by the project under alternate climate change futures.

4.5 Probabilistic Analysis

One drawback to scenario analysis is that scenarios do not address *likelihoods* of the various intrinsic elements required in the modelling process. To circumvent such difficulties, a probabilistic analysis can be employed to permit fundamental types of uncertainty to be directly incorporated into the planning process. In terms of planning for the future operations of the hydro facility, the future streamflows of the river will never exactly match the historic streamflows even without climate change and the patterns of future stream-flows will almost certainly be altered under conditions of a changing climate. To address climate change in the hydro example, there are essentially two forms of probabilistic "styles" that could be readily employed. One approach would solicit opinions based on "expert choice" as to the likelihoods arising from the different scenarios (Morgan & Keith 1995; Wright & Ayton 1992). For example, experts could be used to assign likelihoods to the various stream-flows determined by the streamflow models under each scenario of the previous section. Whenever "experts" are consulted in an EA, it is important for the proponent to specify clearly who these experts are and to define the criteria employed in finding them. In an EA process, the term "expert" could be broadly defined to include outside consultants, local residents and stakeholders in general. The second approach would be to randomly generate representative climate data using stochastic simulation. Stochastic, or Monte Carlo, simulation can be used to estimate probabilities of outputs (i.e. impacts) given probability distributions of model inputs.

A simple illustration of the first probabilistic approach would be to take the four streamflow patterns created by the CSIROM scenarios and initially state that there is an equal likelihood of any one of these scenarios actually occurring. This would be equivalent to assigning a probability of 0.25 to each scenario. The expected future streamflow would then consist of the average of the four equally weighted scenarios. As an illustrative example of the second case, suppose that various hypothetical experts had provided different likelihoods of occurrence to each of the four scenarios with a higher emphasis given to the CSIROM B11 scenario. In this situation, the expected future streamflow would consist of the weighted average of the four streamflow scenarios. These different estimates of expected streamflows are represented in the subsequent two figures in which the expected future streamflow is represented by the larger black line. It can be clearly seen that the expected future streamflow of the first figure corresponds to an average pattern of the four different scenarios, whilst the future streamflow of the second figure more closely follows the streamflow path of the B11 scenario.

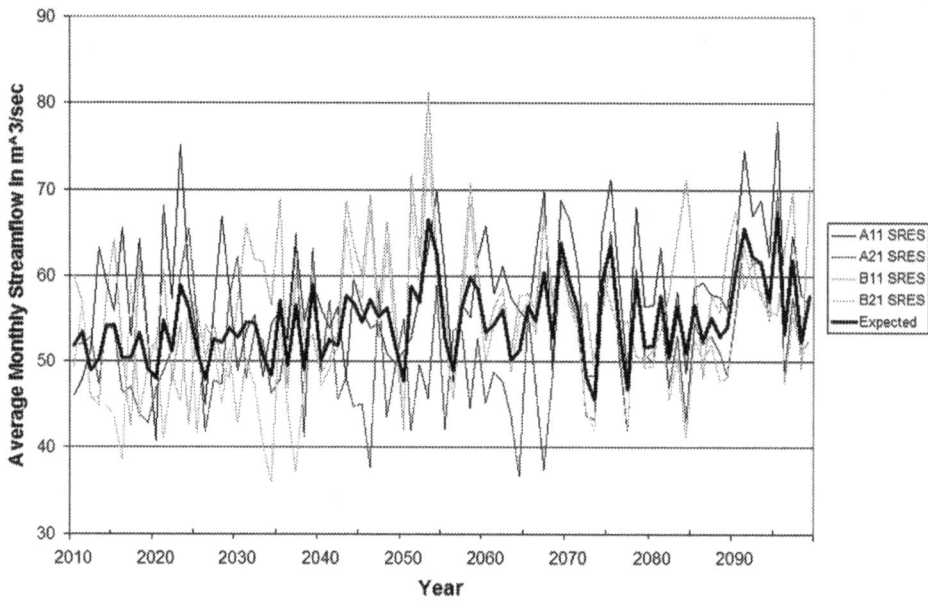

Figure 4.14: Average Monthly Streamflow from 4 Selected CSIROM Scenarios with Equal Weighting Given to Each Scenario

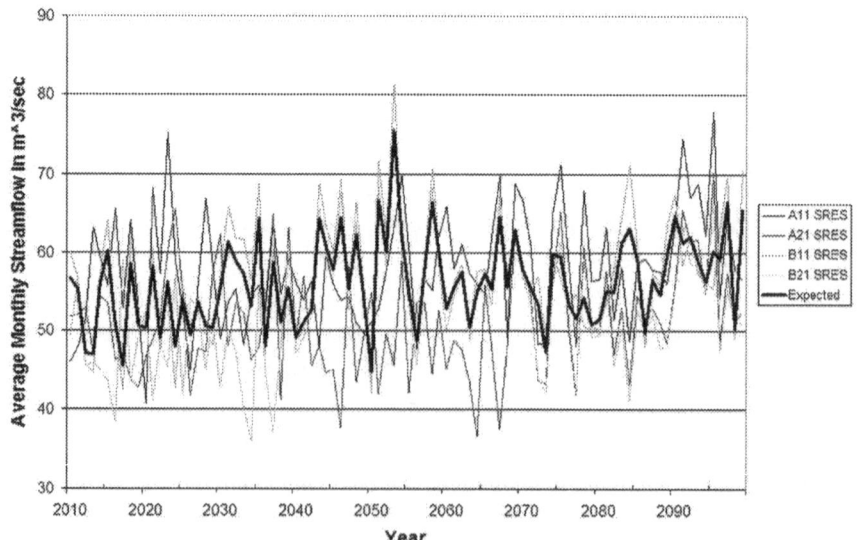

Figure 4.15: Average Monthly Streamflow from 4 Selected CSIROM Scenarios with Higher Weighting Given to the B11 SRES Scenario

An illustration of the second probabilistic approach requires considerably more analytical modelling effort than the first approach. This example will again generate likely future streamflows from each of the CSIROM scenarios, but to do so will require the probabilistic modelling of the monthly values. Using the 30-year time slice of historical streamflow data, empirical probability distributions of the streamflows for each individual month were determined from the data (Evans *et al.* 1993; Law & Kelton 1991) using readily available spreadsheet software (Palisade 2002; Ragsdale 2001). Each of these twelve distributions can be used to generate realistic values for the specific monthly streamflow value by capturing the variability inherent within each month. Such simulations can be easily performed using spreadsheet simulation packages such as *@Risk* (Palisade 2002) or *Crystal Ball* (Decisioneering 2003).

However, distributions resembling historic values would not provide realistic estimates for projecting monthly streamflows into the future time periods, since the climate would be changing. What would be desirable is the ability to simulate future streamflow patterns that would closely resemble the streamflow patterns of the CSIROM scenarios, but somehow capture the streamflow variability that would inherently exist in these future time periods. To achieve this for each of the four scenarios, the values of the future monthly streamflows were generated from the appropriate monthly probability distributions (to capture variability within each month) but these distributions were correlated to the actual data values for that month in the streamflow projections of the respective scenarios (see Palisade 2002 for how such correlated distributions can be created in spreadsheets using @Risk).

The following figure shows 15 random, representative streamflow paths that have been correlated to the CSIROM B21 scenario. Clearly, in order to create a realistic representation of the "true" variability within each scenario, multiple (i.e. several thousand) random streamflow paths would need to be generated.

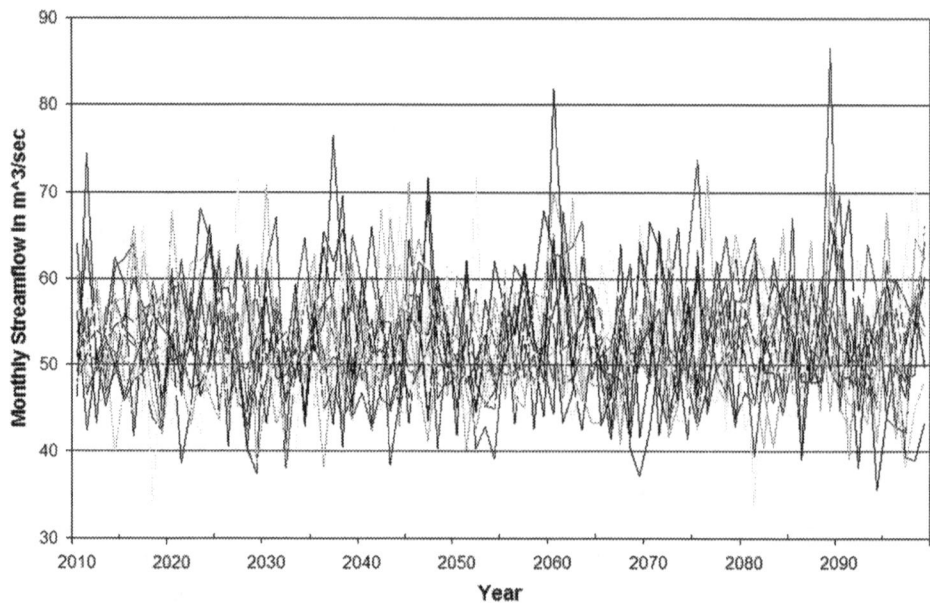

Figure 4.16: Average Monthly Streamflow Rate Produced Each Year by 15
Different Simulation Runs with the Probabilistic Analysis Based Upon the
CSIROM B21 Scenario Data

A picture of this variability appears in the subsequent histograms based
upon a simulation of 10,000 repetitions of each scenario. The first histogram
shows only the streamflows based upon CSIROM B21. The figure indicates
that over the period 2010-2099, the average monthly streamflow is most likely
to lie somewhere between 20 and 120 m³/sec and is almost always less than
200 m³/sec. However, there are extremely rare occurrences when the streamflow exceeds 200 m³/sec and it is even possible for the rate to approach 300
m³/sec. If these rare, high streamflow occurrences corresponded to conditions
of severe flooding, then this histogram might suggest that mitigation and/or
adaptation actions might be requisite in the hydro project.

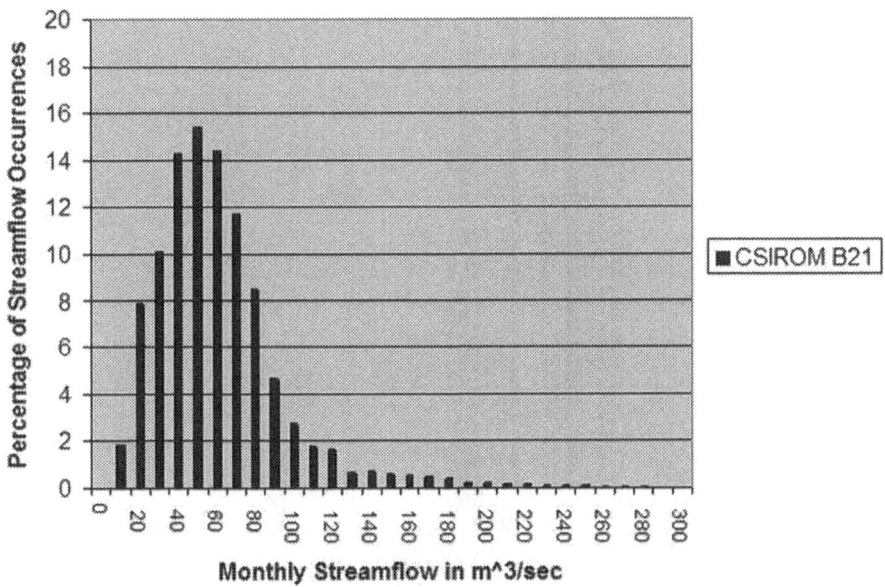

Figure 4.17: Average Monthly Streamflow Likelihoods (2010-2099) with
Probabilistic Data Based Upon the CSIROM B21 Scenario

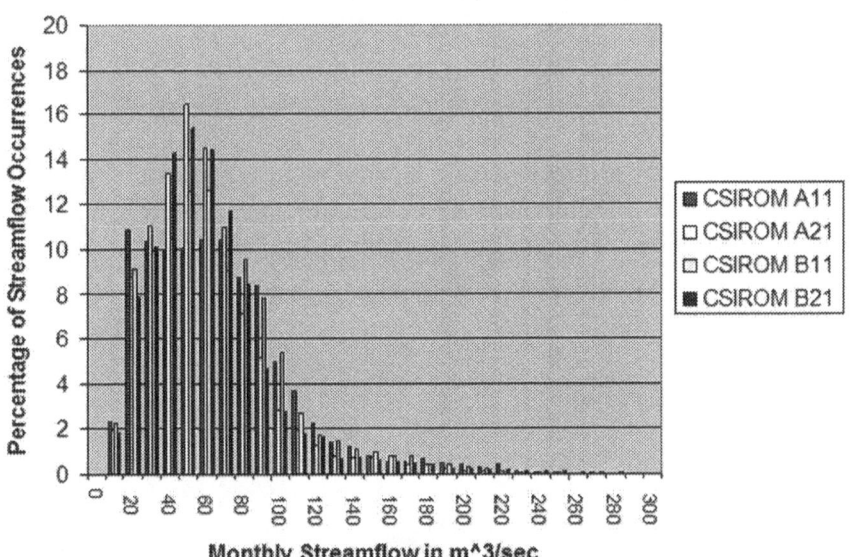

Figure 4.18: Average Monthly Streamflow Likelihoods (2010-2099) with
the Probabilistic Data Based Upon the 4 CSIROM Scenarios

The second histogram shows the likelihood of the future streamflows for the four CSIROM scenarios simultaneously. While there is variability between the different scenarios, the likely streamflow distributions produced by each of the scenarios are relatively similar in shape to each other. Hence, the observations applied to the B21 scenario would similarly hold for each of the other scenarios. Consequently, if these four scenario conditions could actually be considered as representative of all future streamflow conditions, then this probabilistic analysis would indicate that climate change introduces at least a small likelihood of severe flooding conditions. Such information produced during the project's EA might impose a need for adaptation and mitigation responses to be directly included in the design plans for the hydro facility.

While histograms provide an accessible visual representation of this type of probabilistic output data, alternate approaches, such as a table listing the likelihoods for specific streamflow range occurrences, exist that could effectively communicate information to an EA's stakeholders. Although not as visually appealing, such a table would permit an explicit side-by-side comparison of outcome likelihoods under different assumptions of climate change and could also include a column of the historical likelihoods for comparative purposes. It should be noted from the table that the 30 years of historical data available did not show any instances of severe flooding. While the tabular format might be more difficult for some stakeholders to comprehend, the actual information contained in it conveys considerably more analytical details than the other "visual" figures.

Consequently, EA proponents would need to determine the level of technical sophistication of the stakeholders in order to assess the appropriate means of probabilistic communication—perhaps supporting the analysis using a combination of more than one method.

Table 4.8: Percentage Likelihoods of Occurrences of Average Monthly Streamflows over the Period 2010-2099

Streamflow (m^3/sec)	CSIROM A11	CSIROM A21	CSIROM B11	CSIROM B21	Historic
0 to 10	2.36	1.94	2.30	1.82	0.83
10 to 20	10.87	8.79	9.14	7.85	6.11
20 to 30	10.38	11.02	9.93	10.15	10.56
30 to 40	10.02	13.36	10.46	14.32	13.89
40 to 50	10.05	16.43	12.59	15.40	23.06
50 to 60	10.42	14.50	12.65	14.43	20.83
60 to 70	10.45	10.96	9.10	11.74	12.22
70 to 80	8.78	7.12	9.60	8.48	8.61
80 to 90	8.40	5.16	7.77	4.67	2.50
90 to 100	4.97	2.82	5.40	2.74	1.11
100 to 110	3.65	1.94	2.73	1.79	0.28
110 to 120	2.29	1.30	1.72	1.65	0.00
120 to 130	1.42	0.77	1.46	0.70	0.00
130 to 140	1.23	0.74	1.08	0.74	0.00
140 to 150	0.83	0.76	0.98	0.62	0.00
150 to 160	0.56	0.77	0.78	0.55	0.00
160 to 170	0.55	0.40	0.78	0.48	0.00
170 to 180	0.66	0.42	0.44	0.43	0.00
180 to 190	0.52	0.24	0.43	0.25	0.00
190 to 200	0.45	0.14	0.33	0.22	0.00
200 to 250	1.04	0.36	0.30	0.69	0.00
250 to 300	0.10	0.06	0.02	0.27	0.00

Following on from the probabilistic streamflow analysis, the distributional output from the streamflow model would now become the probabilistic input of the energy production model and each series of probabilistic streamflows would generate a corresponding series of energy production. Under an assumption that the design capacity of the hydro facility had been determined using the historical data, the distribution of stochastic energy production outputs arising from the four CSIROM scenarios could be presented visually in the form of a histogram. This figure indicates that under any of the climate change scenarios, the monthly energy production will generally lie between 1.4 and 1.5 Gwh. However, the likelihood of lower energy production is non-negligible and could conceivably be as low as 0.3 Gwh in some months. If the potential for these low energy likelihoods were shown to exist during the EA process, then, as with flooding, it might initiate adaptation and mitigation changes to the design plans of the hydro facility.

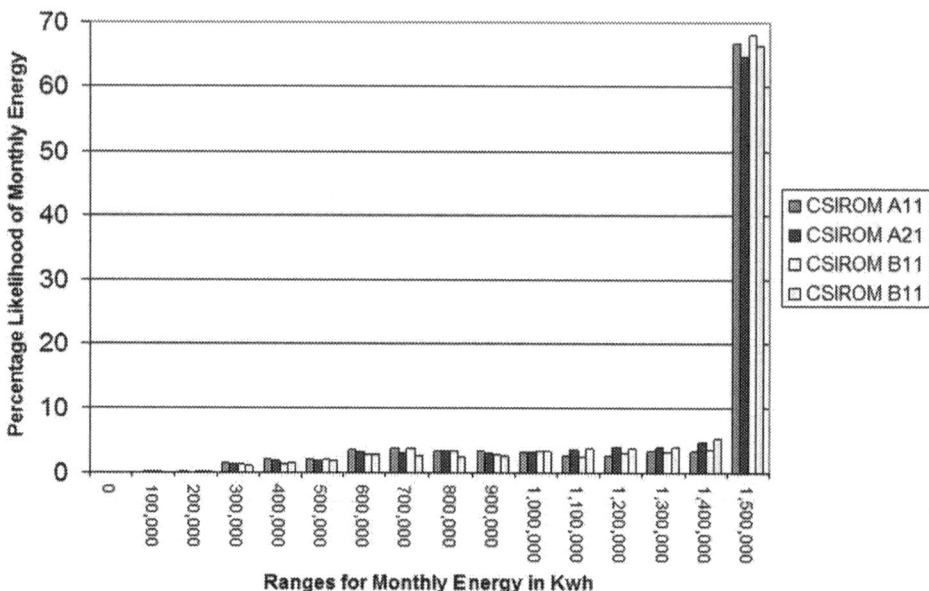

Figure 4.19: Percentage Likelihood of Monthly Energy Production in Hydro Facility with Capacity Based upon Historical Data Under the 4 CSIROM Scenarios

Similar to the probabilistic streamflow case, tabular representations could also be employed for summarizing the probabilistic energy production for the decision-makers and the stakeholders. To permit likelihood comparisons under different climate change assumptions, the following table contains probabilistic data from both the energy and streamflow simulations. For both streamflow and energy, this table shows the expected monthly value (i.e. the average of all of the monthly averages) calculated over the entire time horizon, and the minimum and maximum monthly value (i.e. the minimum/maximum of any of the monthly averages) occurring at any point over the time horizon. Hence, this provides an indication of both the central tendency and the extreme ranges of streamflow and energy. The last row of the table lists the corresponding mean annual reduction in GHG emissions corresponding to the mean energy. While there are numerous other values that might be of interest, a proponent would need to determine what specific probabilistic data would be requisite and appropriate for communicating to the stakeholders of their particular EA.

Table 4.9: Probabilistic Outputs with Capacity Based on Historical Data

	Experiment				
	Historical	CSIROM A11	CSIROM A21	CSIROM B11	CSIROM B21
Mean Average Monthly Flow	48	58	51	57	53
Min Average Monthly Flow	9	2	2	2	5
Max Average Monthly Flow	108	294	290	276	291
Mean Average Monthly Energy	1,302,895	1,240,236	1,251,745	1,261,884	1,269,461
Min Average Monthly Energy	322,621	80,382	82,103	90,920	85,401
Max Average Monthly Energy	1,441,145	1,441,145	1,441,145	1,441,145	1,441,145
Mean Annual GHG Reduction	14,071	13,395	13,519	13,628	13,710

Since the output of simulation tends to be expressed in the form of a probability distribution rather than any single figure, probabilistic analysis can provide challenges as to the appropriate way to convey the meaning of the stochastic impacts determined. Sometimes the impacts could be expressed as average values, while at other times they might be better expressed as ranges or even as visual figures of the probability distributions. In order to establish the most appropriate way to represent their probabilistic output data, EA proponents would need to carefully ascertain the technical sophistication levels of the stakeholders and would perhaps need to communicate the results of the analysis using a combination of several methods.

4.6 Sensitivity Analysis

The final analytical procedure for examining the uncertain conditions in the hydro example is a sensitivity analysis. Sensitivity analysis can provide useful insights into the influence of potential changes in the inputs and can be used for quantifying the impact of variations in these inputs on the values of the outputs. Its main benefit lies in the fact that it produces results that are independent of the correctness of any climate scenario and can be used to identify the climatic variables which have the most significant impacts on a project. That is, if these climate variables are uncertain but could be bounded by an estimated range of probable values, it would be possible to ascertain how vulnerable a potential project design might be to any impacts of the variables. In an EA, this knowledge can be used to guide the direction of an impact study and to highlight the most vulnerable future characteristics in a project which may warrant closer scrutiny.

However, since no singularly best process that specifies the steps for conducting sensitivity studies exists, such exploratory studies necessitate the adoption of numerous "restrictive assumptions" in order for the analysis to become computationally manageable. Consultation with experts and the application of expert judgment proves invaluable in the proper guidance, identification, and selection of the key parameters that could be investigated through sensitivity analysis. For example, expert judgment might be used to predefine changes to the precipitation and evaporation variables that would be input into a hydrological model for determining whether these changes resulted in any significant impacts on a proposed project. At a minimum, the ranges and/or numbers selected for a sensitivity analysis should be sufficient for determining whether or not their impact is significant in the decision-making process. Furthermore, if a sensitivity analysis used values representing pathological extremes (i.e. values beyond the likelihood of ever occurring) and demonstrated that they had absolutely no impacts on a project, then this would indicate that the tested parameter was insignificant and need not be evaluated further.

The (arbitrary) restrictions to be imposed in the following sensitivity analysis of the hydro example are that: (i) only actual historic data will be used in the experimentation; and, (ii) the only style of questioning will be of the form "if input x changes by an amount y, what is its resulting impact on output z?". For the hydro project, five separate "experiments" were conducted involving changes to the historic streamflow rates and the newly altered rates were projected over the time horizon, 2010-2099. In these experiments, it has been assumed that the design capacity of the facility was determined based upon the existing historical streamflow data. Hence, this type experimentation will effectively consider what could potentially happen to a currently designed facility in the face of varying rates of streamflow. The specific impacts of these streamflow changes were measured by the resulting levels of energy production, GHG emissions reductions, and "extreme" streamflow rate potentials.

The five changes to the historic streamflow rates involved: (i) a 15% increase to the mean streamflow rate with no change to the standard deviation of the flow rate; (ii) a 15% decrease to the mean streamflow rate with no change to the standard deviation of the flow rate; (iii) a 15% increase in the mean streamflow rate accompanied by a 25% increase to the standard deviation of the flow rate; (iv) a 15% decrease to the mean streamflow rate accompanied by a 25% increase to the standard deviation of the flow rate; and, (v) no change to the mean streamflow rate but with a 25% increase to the standard deviation of the flow rate. These five experiments are summarized in the Table 4.10 and their impacts on the streamflow rates, together with their corresponding impacts on energy production and GHG emissions, can be seen in Figure 4.20 and Figure 4.21.

Table 4.10: Sensitivity Analysis Experiments Performed

Experiment	Change to Mean Historic Flow Rate	Change to Standard Deviation of Historic Flow Rate
Experiment 1	15%	0%
Experiment 2	-15%	0%
Experiment 3	15%	25%
Experiment 4	-15%	25%
Experiment 5	0%	25%

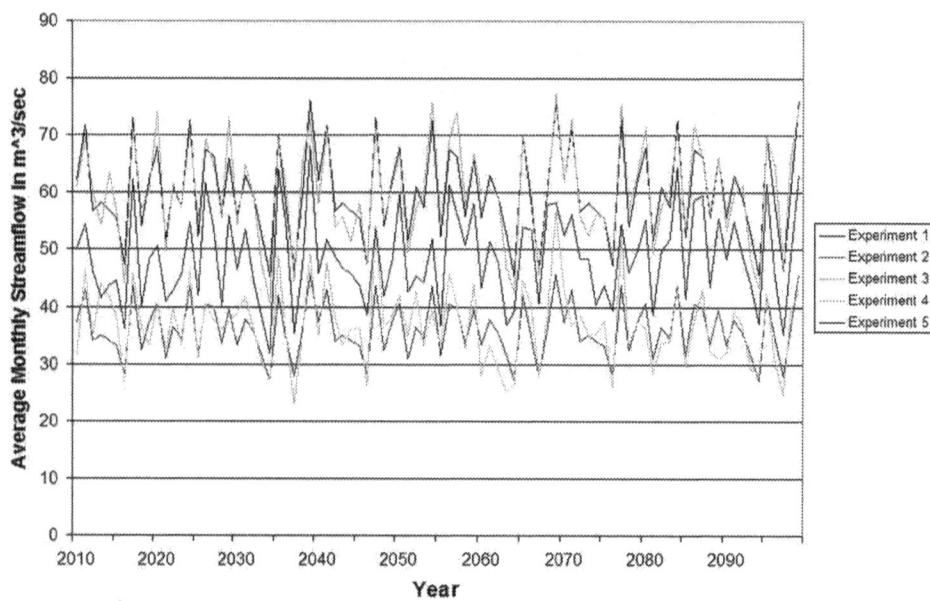

Figure 4.20: Average Monthly Streamflows for 5 Variations on Projections of Historical Flow Rate Data

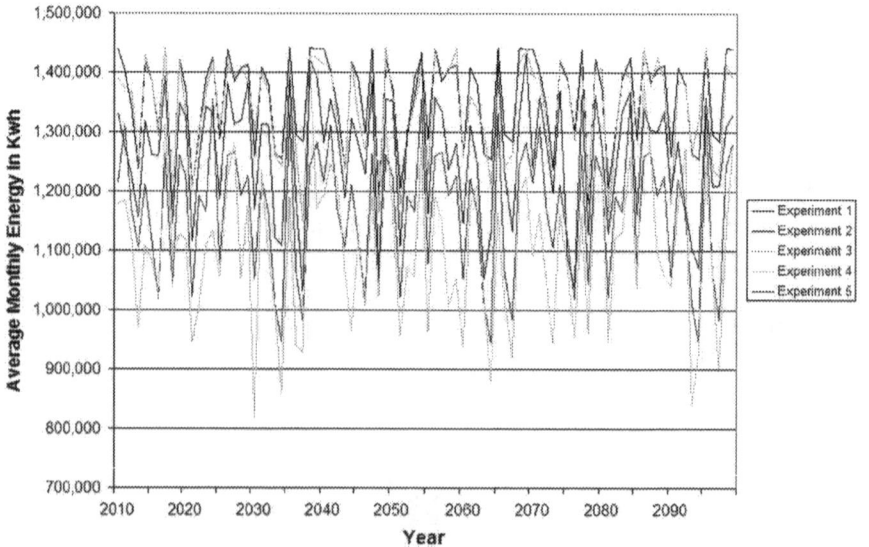

Figure 4.21: Average Monthly Energy for 5 Variations on Projections of Historical Flow Rate Data

As with the probabilistic analysis, there are a wide variety of approaches involving combinations of graphical and tabular methods that could be used to communicate the results and uncertainties produced in a sensitivity analysis. The proponent would need to determine whether the method(s) employed sufficiently captured the nature of the changes and impacts uncovered in the analysis or whether other presentation methods might be more helpful to support an effective analysis. The approach selected by a proponent would necessarily have to depend upon the complexity of the output, the type of information that needed to be communicated, and the level of technical sophistication of the interested stakeholders. A tabular summary of the experimentation is shown below in which the resulting values of the mean monthly energy, its corresponding reduction in GHG emissions, and largest (or extreme) monthly streamflow over the entire horizon are provided.

Table 4.11: Sensitivity of Historical Values to 5 Changes in Historical Streamflow Rates

	Historical	Experiment 1	Experiment 2	Experiment 3	Experiment 4	Experiment 5
Change to Mean	0%	15%	-15%	15%	-15%	0%
Change to STD	0%	0%	0%	25%	25%	25%
Average Monthly Energy (Kwh)	1,302,895	1,355,618	1,167,060	1,341,103	1,101,460	1,254,385
Annual GHG Reduction (tonnes CO2)	14,054	14,641	12,604	14,484	11,896	13,547
Maximum Monthly Average Flow (m^3/sec)	108	135	81	146	137	143

It then becomes necessary to determine if any of these impacts might be considered significant. An analysis of the table would identify that, in this example, the variability of the average energy and GHG emissions levels seems to be more significantly impacted by changes to the mean streamflow rate than by changes to the standard deviation of the flow rate, and that the possibility of the occurrence of extremely streamflow events seems to be more impacted by larger increases to standard deviation of the streamflow rate than by changes to the mean flow rate. Therefore, if realistic bounds on climate induced changes to streamflows could be established that indicated a high likelihood of these types of occurrence, then, based upon this information, the hydro project could be redesigned accordingly to accommodate any potential impacts that might arise from such uncertainties.

Knowledge of the sensitivities in the EA may direct further impact/sensitivity studies to evaluate potential design changes that could be incorporated into the hydro project in order to adapt to or mitigate against any resulting vulnerabilities. In fact, sensitivity analysis could be applied as a "semi-quantitative" screening tool to virtually all analytical cases. A proponent could employ expert judgment to select a few key climate variables believed to have a major impact on the project and use sensitivity testing to provide an indication of whether more detailed climate change evaluation was warranted. In the hydro example, the proponent might want to determine what other sensitivities might prove essential for effective decision-making. For instance, if it could be determined what level of energy production were necessary in order for the investment in the hydro project to prove economically worthwhile, it might be necessary to determine the thresholds of several key climate variables (i.e. those impacting future streamflows) that subjected the project to "make or break" economic sensitivities. A subsequent examination of the vulnerability of the project to the uncertainties in these climate variables would then be requisite. Hence, sensitivity analysis can be used to identify the critical climate variables that could significantly impact the project and which project design vulnerabilities to focus upon given the inherent uncertainties in a predicting the changes to these climate variables.

4.7 Effects of Climate Change on Project Impacts

Thus far, the hydro example has been used to incorporate the uncertainties relating to (i) the climate change impacts *on* the project (i.e. changing streamflows) and (ii) the project's implications *for* climate change (i.e. greenhouse gas emissions). In addition to these impacts, it is also necessary for EAs to consider the effects of climate change on the impacts of the project. Specifically, an EA must determine the impacts on all of the affected *valued environmental components* (VECs) such as wetlands, groundwater, fisheries, and jobs. For the hydro project, the assessment of VECs is restricted to an examination of the impacts on flooding potential and the downstream fisheries. Much of the uncertainty information for these components can be drawn from the earlier analyses of uncertainty, while additional sources of data are required to address certain other components. Furthermore, the three techniques for examining uncertainty can all be employed, either individually or in combination, to address these uncertainties, so all of the previous discussions are equally applicable for analyzing and addressing the impacts on VECs.

4.7.1 Flooding

Suppose that an additional background study of the current daily streamflow conditions had indicated that rates in the range of 200 to 250 m^3/sec result in low levels of flooding damage, while rates in excess of 250 m^3/sec produce moderate levels of damage. In addition, historical records also indicated that low levels of flooding have occurred with an annual probability of 0.08 and that moderate flooding has occurred with an annual probability of 0.02. In order to assess the flooding impacts arising from climate change and the hydro project, the EA would need to determine any changes to the likelihood of flooding that might occur and also any changes in the extent of damages attributable to this flooding.

In the first step of this analysis, the proponent might wish to ascertain exactly how vulnerable the region might be to flooding and decide that a sensitivity analysis would be useful. In order to ascertain which key parameters to focus upon in this analysis and to simultaneously screen out any potentially irrelevant factors, the proponent would need to consult with experts on this topic. Since it is the extreme daily flow values that create the flooding damage, suppose that this expert judgment determined that the key sensitivities to investigate revolved around the variability in streamflow rates. Consequently, a sensitivity analysis could be conducted measuring the changes to flooding damage likelihoods based upon changes to the standard deviation of the streamflow rates (changes which thereby impact the likelihood of both higher and lower extreme streamflow values). A table using illustrative changes to the standard deviation and showing the outcome of this sensitivity analysis appears below.

Table 4.12: Sensitivity Analysis for Flooding: Annual Probability of Damage

Annual Probability of Damage

Damage	Historic	With Increase in S.D. by:			
		15%	25%	22%	12%
Low	0.08	0.14	0.22	0.20	0.08
Medium	0.02	0.05	0.11	0.08	0.04

This table demonstrates, for this example, that as long as any increase in streamflow variability remained lower than 12%, the likelihood of "low" flooding damage remains unchanged from the current conditions. For increases beyond 12%, the risk of low flooding damage increases, with the likelihood increasing by a factor of almost 3 for a change of 25%. Thus, it would appear that the region might prove susceptible to increased risks of low flooding damages if a changing climate introduced significant variability to the streamflow rates. Similar observations can also be made with respect to the risks of "medium" flooding damages. With a 15% change in variability, the risk of medium flooding damage doubles and with a change of 25%, the risk increases by a factor of 5. Suppose that based upon further expert consultation, it was determined that changes to the standard deviation of the streamflow rates within these ranges would be entirely possible under a changing climate. Then, such a sensitivity analysis conducted during an EA would indicate that this region had the potential to experience considerable increases in flooding problems under climate change and that flood control actions should be considered in the EA for the project.

Subsequent to the findings from the sensitivity analysis, the proponent may wish to determine the impacts of climate change on flooding in more detail using the scenario and probabilistic approaches considered previously. Since any flooding along the river would occur during periods of high streamflow events, output from the probabilistic streamflow analysis would be essential to this phase of the analysis. Daily streamflow data would be required for this analysis, since flooding would depend on the daily peak streamflow as opposed to the averaged monthly values which were used in the previous sections.

However, the daily data created during the probabilistic analysis to produce the monthly values could be used to assess the likelihoods of flooding. The likelihoods of daily streamflow, created through a combination of probabilistic analysis with the four CSIROM scenarios, appears in the following table together with the historical streamflow information.

Table 4.13: **Scenario and Probabilistic Analysis of Flooding**

Probabilities of Occurrences for Daily Streamflows, 2010-2099

Streamflow (m^3/sec)	A11C	A21C	B11C	B21	Historic
200 to 210	0.042	0.009	0.026	0.017	0.025
210 to 220	0.033	0.010	0.020	0.017	0.021
220 to 230	0.025	0.004	0.010	0.013	0.015
230 to 240	0.017	0.004	0.010	0.011	0.013
240 to 250	0.018	0.009	0.000	0.011	0.006
250 to 260	0.013	0.002	0.006	0.007	0.005
260 to 270	0.014	0.002	0.001	0.009	0.006
270 to 280	0.010	0.001	0.001	0.008	0.003
280 to 290	0.012	0.000	0.000	0.002	0.001
>290	0.019	0.002	0.005	0.004	0.005

An examination of the probabilities for flooding indicates that under one scenario of climate change (A11), the risk of both low and moderate flooding damage increases substantially over the current conditions, while under another scenario (A21) the converse is true. The patterns of flooding likelihoods under the remaining climate scenarios (B11, B21) remain approximately the same as those currently experienced in the region. Hence, if these climate change scenarios could be considered representative of the future climate in the region, such an analysis might indicate the potential for significant damages and that certain flooding mitigation strategies might need to be introduced into the hydro project's EA.

4.7.2 Fisheries

For illustrative purposes, the second VEC to be considered as part of the hypothetical hydro example is an analysis of the climate impacts on the fisheries immediately downstream of the proposed site. Such an analysis would need to explore not only the climate change impacts on the survivability of the fish species in the river, but also the resulting impacts on recreational fisheries and related tourism in the area. Under current conditions, the river has approximately 1,400 m² of primary fish habitat and includes a significant sport fishery, primarily for the warm water species. The habitat can be divided into two primary components: (i) an area of shallow fast moving water predominantly containing warm water game species; and, (ii) several deep basin areas inhabited by cold water species.

The input of a fisheries expert would be needed in this phase of a study in order to determine, and set values for, any key parameters and to assess the results. Suppose that this expert determined that the primary impacts of climate change and the hydro facility on the fisheries would be due to changes in: (i) the streamflow rates of the river; and, (ii) the temperature of the water. Since these impacts can involve climate change, as before, the methods of scenario analysis, probabilistic analysis and sensitivity analysis can be employed to assess the uncertainties.

The potential impacts on the survivability of the fish species could first be assessed using a sensitivity analysis that focused on changes to the two key parameters identified by the expert; streamflow rate and water temperature. Experts would also need to specify the possible ranges in these parameters that could result from climate change. The fisheries expert could then ascertain the impacts of combinations of these changes in both streamflow rates and water temperature, e.g. streamflow is changed by +/-10% and the temperature is changed by +/-1.0°C. Suppose that the expert's opinions on these combinations appears in the table below, with the terms "significant" and "insignificant" properly defined in the EA.

Table 4.14: Sensitivity Analysis of Fisheries

Water Temp Changes by:	Streamflow Changes by:	
	-10%	+ 10%
-1.0C	Significant loss of warm water species	Minor loss loss of warm water species
+1.0C	Significant loss of warm and cold water species	Insignificant effects

Hence in this hypothetical example, based upon this sensitivity analysis, there appears to be the potential for significant losses to occur in the fishery under certain circumstances. Suppose it was determined that the sensitivity values used for the changes to temperature and streamflow could be considered well within the possible variable ranges that might occur under conditions of a changing climate. Because of this, the possible impacts from climate change and its uncertainties on species survivability might then be further assessed using a combination of scenario and probabilistic analysis. For example, different possible futures of streamflow scenarios and temperature scenarios could be shown to the fisheries expert who could provide advice on which scenarios might result in a significant loss of the warm water species.

4.7.3 Recreational Fishing and Tourism

The final phase of the fisheries study would be to evaluate the potential impacts of both climate change and the project on recreational fisheries and tourism within the area. In particular, the key consideration would be the impacts on the warm water species, since this represents the primary draw for tourism. Hence, this portion of the study requires two stages of assessment: (i) the specific impacts on the warm water species; and, (ii) the threshold changes leading to a collapse of the warm water species on recreational fishing and tourism. This two stage assessment could be incorporated into a sensitivity analysis.

In stage (i), a fisheries expert would again be needed to determine what changes in climate variables (precipitation, temperature, streamflow, water temperature) would cause different levels of impacts to the warm water species. This would require that one would need to judge the answer against various possibilities of climate change. From the previous sensitivity study, it was shown that the warm water species would decrease under conditions of either reduced streamflow rates or reduced water temperature, and that no combination of temperature/streamflow considered ever resulted in an increase to the warm water populations.

In stage (ii), the specific threshold values for the key parameters from stage (i) that might lead to a collapse of the warm water species would need to be determined. It has been demonstrated that the warm water species would encounter *significant* losses under the combined conditions of a 1.0°C reduction in temperature and a 10% reduction in streamflow rate. Suppose that, after subsequent expert consultation, it was determined that the threshold values leading to a *complete* collapse of the recreational fishery were a combination of a 1.5°C decrease to temperature and a 12% decrease to the streamflow rate. If these potential changes to temperature and streamflow fell within reasonable ranges of what might occur with climate change, then it might be conceivable for the warm water fish species to face extinction within the region under certain foreseeable future conditions.

Once the potential impacts from the two-stage sensitivity analysis have been determined, their potential impacts on recreation and tourism could be estimated. Suppose that under current conditions, there are 675 user-days of recreational fishing, consisting of 475 user-days attributable to local residents and 200 user-days attributable to tourists. The effect of climate change on this tourism and recreation industry would be dependent upon the effect on the recreational fisheries as determined by the sensitivity, and possibly other, analyses. Since, in this example, the sensitivity analysis shows that under no situations would the recreational warm water fishery expand, the current 675 user-days represent the upper bound for the recreational fishery. However, combinations of applicable climate conditions indicated a loss in the warm water fisheries, which would result in a loss to the recreation and tourism industry, and, some combinations of climate conditions could lead to a complete collapse of the tourism and recreational fishing industry, resulting in a loss of the entire 675 user-days. Such findings would need to be explained in the environmental assessment.

Hence, using the hypothetical hydro example, this section and the previous two sections have demonstrated how the three analytical techniques previously introduced might be modified and employed, either individually or in combination, to analyze and address various uncertainties and impacts on VECs. Such a series of analyses on the region's flooding, fisheries and tourism would be needed within the hydro project's EA in order to fully express and assess the potential impacts.

4.8 Choice of Method

The hydro project presented in this chapter has been used to illustrate how three common methods can be used individually and in combinations to analyze several typical impacts that are commonly studied in EAs (i.e. energy production, GHG emissions, flooding, fisheries, recreational fishing and tourism). The example demonstrated that an approach combining scenario analysis with probabilistic analysis provided an appropriate method for understanding the effects of climate change uncertainties on streamflow and energy production. In this combined approach, the scenarios were used to provide quantitative estimates of the climate variables that served as the data inputs to the stochastic models of streamflow and energy. Conversely, the sensitivity analysis method would be more appropriate, at least as a first step, for addressing the uncertainties in the potential impacts on the downstream fishery, since the link between streamflow and the fishery was more complex and less understood.

The purpose of this hypothetical example has been to demonstrate which analytical methods can be most appropriate for incorporating the different effects of climate change uncertainties into *this* particular setting. While there can be no definitive rule for determining when to select one method over another for addressing climate change uncertainties, this section discusses factors that should be considered.

A key factor in the choice of method is based upon the nature of the measurability of the data. Namely, the applicability of a method can depend upon whether the values it requires are well-defined and quantitatively measurable or are only ill-defined and qualitatively measurable. In addition, the level of

difficulty in the use of a method should play a major role in the appropriateness of its selection, in that methods requiring significant use of resources (i.e.
expertise, time, data, cost) should not be used to study less important impacts.
Furthermore, methods requiring significant modelling effort necessitate the
existence of a clearly developed level of understanding about the relationships
between climate change and the impact (i.e. the existence of well-developed
analytical models).

Probabilistic methods require the existence of both well-developed models
and well-defined, quantitative data for input into these models. Probabilistic
analysis also requires considerable technical expertise and effort on the part of
the user. Hence, in order to justify the significant development time, computational efforts, and economic costs needed for probabilistic studies, the
importance of the impact being studied would need to be high. Clearly, if an
impact can only be measured qualitatively or cannot be defined probabilistically, then probabilistic methods cannot be employed.

A *sensitivity analysis* can produce a good first-step in most analyses, since it
can essentially be applied as an analytical "screening device" (as described earlier) in virtually all cases. If the impact being studied is of relatively minor
importance in the EA, then additional analysis would not be required. Furthermore, given its wide applicability, sensitivity analysis can provide the *only*
choice when it is not possible to employ the other methods. Thus, sensitivity
analysis can prove to be the appropriate choice when either the models and
data required for the other two methods are not readily accessible or when the
importance of the studied impact to the project is relatively low.

Due to the combinatorial explosion in computation time, sensitivity analysis cannot be performed on anything more than a very narrow set of parameters. To counteract such computational difficulties, the next methodological
progression is toward a *scenario analysis*. Scenario analyses require more extensive computational efforts, better developed models, and a better quality of
quantitative data than sensitivity analyses. Hence, a scenario analysis could
prove to be the most appropriate choice when the quality of the model and
quantitative data availability is reasonably substantial, and when the impact
being studied is of more than minor importance to the EA. This will often be
the case for impacts directly linked to climate change variables because of the
development of climate change scenarios that has already been done.

Therefore, two of the major factors that should influence the choice of the
analytical approach are: (i) the importance to the project of the specific impact
being studied and the importance of the information resulting from the analy-

sis; and, (ii) the quality of the models that are available to study the impact and the quality of the quantitative data that is available for use in the models.

Based upon this identification, one possible framework for selecting which method to apply appears in the table below.

Table 4.15: Choice of Method

Importance	Model and Data Availability		
	Poor	Fair	Excellent
Low	None	Sensitivity	Sensitivity or Scenario
Medium	Sensitivity	Scenario	Scenario
High	Sensitivity	Scenario	Scenario and Probabilistic

Depending upon the particular EA, each of the three analytic methods, or various sequences and combinations of them and the proponent, in consultation with experts and stakeholders, would need to justify the choice of method(s) to employ.

4.9 Conclusion

While many proponents have openly recognized the potentially catastrophic consequences associated with future climate change, several studies have indicated that the corresponding impacts and uncertainties of a changing climate have not been adequately addressed in EAs. A major difficulty facing many project EAs is determining how changing climatic conditions can impact the project and how to effectively incorporate any resulting uncertainties into this analysis. Given the lack of consensus in scientific predictions of the future climate, many proponents have either tended to disregard the impacts of climate change or have maintained that it would be too complex to account for the impacts of the uncertainties in their EAs. Some proponents believe that the effort required in conducting a climate change assessment will not yield practical results that improve a project design (i.e. that the level of effort required is not proportional to the added design value). However, since there are many ways that a project could be designed to adapt to a changing climate, there are legal obligations for proponents to consider the potential impacts and uncertainties from climate change in their EAs. In spite of the uncertainties, proponents must recognize that they need to actually need to be providing expert judgments at a particular point in time given the information currently available, even when those judgments might contain considerable elements of subjectivity (Moss & Schneider 2000; Ravetz 1986).

The incorporation of climate change uncertainties into an EA is neither a straightforward process nor easily accomplished. Using an example based on a proposed hydro-electric facility in Northern Ontario, this chapter considered several different methods that could be used to address the uncertainties arising from climate change in impact assessments. Proponents would need to adapt these (or other) methods to the situations faced within their specific EAs in order to incorporate the relevant climate uncertainties into their own analyses.

Once uncertainties have been determined, it becomes essential to communicate the results to the various stakeholders in the EA. How such complexities should be effectively communicated to stakeholders with differing levels of technical sophistication is addressed in the next chapter.

5

Communication of Climate Change Uncertainty

5.1 Introduction

While a considerable body of literature has been compiled in the development of plans and methods for adapting to the impacts from climate change, a similar accounting for climate change impacts has not been included within the structure of EAs, let alone for the incorporation of any climate change uncertainties into the EA process. However, there can be many ways in which these uncertainties can be addressed, lessened, and/or adapted to at the project level and, therefore, there is a legislated requirement (Canadian Environmental Assessment Act, Sections 2 & 16) for these approaches to be readily considered in EAs. Whereas the previous chapter demonstrated how climate change impacts could be *incorporated* into the EA analysis, this chapter examines how the resulting climate change uncertainties could be effectively *communicated* to the disparate stakeholders of the EA process.

The extent of any mitigation and adaptation activities proposed during the EA process depends upon the impact uncertainties determined during the analysis of the project, which in turn depends inherently upon the range of uncertainties and assumptions made regarding future climate change. For instance, in the hydroelectric project example in the previous chapter, the uncertain impacts from climate change directly created uncertainties about future streamflows, which in turn created uncertainties regarding future energy production, future flooding potential, and the future of the downstream fisheries. While little may be done to proactively prevent the uncertain future streamflows caused by a changing climate, the project, itself, could be modified, designed and operated to reduce the effects of, and vulnerabilities to, the remaining uncertainties. For example, the hydro project could be modified to include an accompanying reservoir that would be sized with an outflow release strategy designed to reduce the range of downstream flooding risks. However, any strategies designed to mitigate and adapt to climate change uncertainties require that these uncertainties be sufficiently addressed during the EA and sufficiently revealed to all stakeholders in the communication and elucidation process.

5.2 Role of Stakeholders

One of the most important aspects of EA processes in Canada is the involvement of the various stakeholders. The wide-ranging intersubjectivity provided by the stakeholders contributes a considerable source of strength to the whole systems approach of the EA. For this involvement to be both meaningful and effective, the appropriate information derived from the impact studies has to be clearly communicated to, and understood by, a wide range of audiences. This type of broad communication would be unnecessary when dealing with traditional scientific information, destined to be used within a well-defined community of scientific colleagues sharing the same family of techniques and the same goals. The communication requirements in an EA are completely different, since there are numerous fields in public policy-relevant areas whose status can be violently contested by many different stakeholders each legitimately possessing their own agendas, valuations, and perceptions. The great lesson that scientists in EA processes have learned is that these stakeholders will not defer to the research scientist's "special vision" of problems and solutions. Furthermore, the pragmatic reliability of scientific information cannot be tightly bound to its status as "correct" knowledge in public policy, since what is effectively scientific knowledge at any one time is liable to subsequent revision when viewed with the wisdom of hindsight. For better or for worse, stakeholders in EAs demand an open forum for the airing of their own opinions and their management of environmental uncertainties, and their criteria for the quality of information, can prove to be quite different from those of scientists (while also differing considerably amongst themselves). In such fluid situations, where opinions, knowledge, and ignorance can become so intertwined that traditional categorizations prove scarcely applicable, intellectual discipline and practical guidance in the communication of the degrees of uncertainty (and corresponding levels of ignorance) can help considerably in the clarification of the debate.

Depending upon the data, models, and assumptions used to estimate the impacts, the stakeholders will possess different levels of "trust" in the results. In policy processes for decisions on environmental problems where scientific results have frequently been obtained from computer models, these results can prove to be especially controversial. Given the contentious debates within and outside scientific circles about the degree of changes that could be caused by GHG emissions, there may be considerable differences among stakeholders in their acceptance of any results from a scenario analysis. For example, a study based on a certain set of scenarios might conclude that the annual risk of downstream flooding as a result of climate change would increase insignificantly from 0.001 to 0.002. However, the downstream residents may believe that this range is far too conservative since more extreme or "surprise" scenarios had not been considered in the analysis. Somehow this type of apprehension and uncertainty regarding the climate change uncertainties and data reliability has to be appropriately conveyed to the users of the EA analysis.

Hence, what seems imperative is the ability to examine and communicate information that will help the stakeholders and decision-makers assess their beliefs in the estimates and to be able to readily discuss these as an integral part of the EA process. For climate change outcomes, it becomes clear that in the absence of more "classical" procedures for the scientific validation behind the evaluation of results, the guiding principle must assume the form of some type of quality assurance standard for the communication of the information. Operating under this principle, if there are subsequent queries about the quality of the scientific information in a policy issue, then this assessment of quality can serve to establish some explicit common ground for an orderly discourse between the stakeholders. A well-designed EA process could, therefore, serve to increase consensus, and also enhance awareness, on the uncertainty and quality of information, without embroiling its users in contentious debates over peripheral issues.

Irrespective of the degree of stakeholder consensus, the communication of climate change uncertainties can prove to be particularly complicated, and various studies have made different recommendations on what issues and factors should be considered in order for the stakeholders to attain this good understanding of the uncertainties and their implications–and how this information can be effectively communicate to them. What is needed for effective policy making is a form of analysis framework or method for communicating the uncertainties that serves as the standard guide for the presentation of analytical results in any particular project EA.

5.3 Methods for Communicating Uncertainty

Most environmental systems are representative of highly complex systems that are complicated by numerous inherent uncertainties and by a plurality of stakeholder perspectives (Chociolko 1995; Funtowicz & Ravetz 1999). Although science has long been viewed as providing an objective source for definitive answers, it has now become recognized that science is seldom absolute when applied to the complex systems actually addressed in public policy (Funtowicz & Ravetz 1984, 1990, 1999). Even when scientific methods prove sufficient to establish ranges and likelihoods of the uncertainties in environmental systems, these expressions neither communicate any indication as to the _quality_ of these estimates nor do they permit ready comparisons to be made between disparate estimates and methods. When uncertainty and variation are understated or suppressed in the communication of public policy, a false impression that "everything is known" becomes pervasive (Hammitt & Shlyakhter 1999).

To bridge this deficiency requires the development of a framework for conveying the gaps in existing knowledge by capturing and expressing the inexactness, unreliability, and borders-of-ignorance that are present in the scientific estimates of uncertainty. Researchers and practitioners have examined numerous different approaches for processing and communicating the uncertainties inherent in complex systems. These practices have ranged from the very quantitative to the very qualitative, and from the mathematical to the entirely graphical. If a universally applicable method for presenting climate change uncertainties in EAs could be determined, then the actual communication of the uncertainties would clearly be dependent upon both this specific method and the characteristics of the results actually produced by it. In this section, several possible approaches are briefly summarized and their suitability for the communication of climate change uncertainties in project EAs is considered.

5.3.1 IPCC Methods

In preparation for the IPCC's Third Assessment Report (TAR), Moss & Schneider (2000) assessed several means for characterizing climate change uncertainties and prepared a guidance paper for use by all TAR authors. Noting the need for a consistent approach, Moss & Schneider (2000) proposed not only a general process for assessing the uncertainties, but also several specific tools that could be used to communicate them. Six general steps were recommended for assessing the uncertainties in TAR. These steps can be summarized as follows: (1) identify the most important factors (e.g. processes, variables, parameters, data, and interdependencies); (2) document ranges and distributions in the literature, and distinguish findings that have been well established through observations and tested theory from those that are not so established; (3) make an initial determination of the appropriate level of precision (i.e. given the state of knowledge, are only qualitative estimate possible, or is quantification possible, and if so, to how many significant digits); (4) characterize the distribution of values that a parameter, variable or outcome may take–what is the range, what sort of values could be considered "outliers", what portion of the range corresponds to a specified confidence interval (e.g. 90%), what is the general shape of the distribution, and, if appropriate, what are the central tendency or "best guess" estimates; (5) rate and describe the state of scientific information on which the conclusions and/or estimates are based, reflecting both the type/amount of evidence and their level of peer acceptance/consensus; and, (6) prepare a "traceable account" of how the estimates were constructed.

One important recommendation was that extreme care has to be taken to avoid vague and/or overly broad statements that prove difficult to either support or refute. It was further recognized that because words used as descriptors can hold very different meanings to different stakeholders, that verbal descriptions of scientific information must be calibrated consistently. Hence, for communicating uncertainties in the TAR report, it was prescribed that verbal confidence descriptors be translated according to following quantification system.

Table 5.1: Quantification of Verbal Descriptions in TAR Report

	LIKELIHOOD RANGES	
VERBAL DESCRIPTOR	FROM	TO
Very High Confidence	0.95	1.00
High Confidence	0.67	0.95
Medium Confidence	0.33	0.67
Low Confidence	0.05	0.33
Very Low Confidence	0.00	0.05

Moss & Schneider (2000) recognized that it is seldom possible to achieve this translational precision in the adoption of such a quantification scheme. Therefore, in order to permit a more consistent qualitative communication of the state of knowledge, they proposed that the more general verbal descriptions used in the following table be applied when more quantifiable details were not readily available.

Table 5.2: Verbal Descriptions for Qualitative Communication

		AMOUNT OF EVIDENCE	
		Low	High
LEVEL OF AGREEMENT &/OR CONSENSUS	Low	Speculative	Competing Explanations
	High	Established but Incomplete	Well Established

This qualitative classification scheme could provide a valuable basis for communication in the TAR by using a vocabulary which is straightforward for users to actually understand and implement. Unfortunately, it necessitates that some highly subjective judgment be applied in order to partition specific categories into their "high" and "low" levels (i.e. where does the boundary/threshold between these two categories actually lie?), and no universally available process can be supplied to accomplish this task.

Finally, Moss & Schneider (2000) suggested that ancillary approaches for characterizing the uncertainties associated with key findings can increase the clarity of the conclusions achieved in the TAR. They concluded that the communication of these uncertainties can be significantly improved using graphical representations of the results, and provided two explicit suggestions for this expression using (i) Tukey's box-and-whisker plots and (ii) radar/snowflake charts. However, the two methods provided are not useful for representing situations involving multiple attributes, so they advised that any selection of graphical presentation for communicating uncertain quantitative information be left to the specific user.

Consequently, while the information for communicating uncertainties supplied by the IPCC is applicable and useful at the macro level, it only provides subjective details and these cannot be directly transformed into a specific framework for clearly communicating climate change uncertainties to the different stakeholders of an EA.

5.3.2 NUSAP

Recognizing the inexactness, unreliability, and borders-of-ignorance prevalent in scientific estimates of uncertainty, Funtowicz and Ravetz (1990) proposed the *NUSAP* formalism for communicating both quantitative and qualitative evaluations of the prevailing uncertainty in environmental and social research. *NUSAP* is an acronym for *N*umeral, *U*nit, *S*pread, *A*ssessment and *P*edigree. The *Numeral* refers either to a number or to a set of elements and relations that expresses magnitude, intervals, rankings or even non-numerical descriptions such "small", "medium" and "large". Numeral is the most quantitative element within NUSAP. *Unit* refers to the scale of what is being measured, such as "grams", "watts", "speed", and "number of different species". The *Spread* expresses the range of uncertainty or inexactness in the numeral estimate and could be expressed as: a variance or standard deviation; a specific confidence interval; a percentile estimate; lying within a factor of "n" of the estimate; lying within a logarithmic range of the estimate, or; any other form of entropy measure from information theory. *Assessment* provides an estimate of the reliability associated with the information. It can be expressed more quantitatively through confidence limits and significance levels or more qualitatively using less formal descriptors for expressing degrees of optimism in the estimates through the use of such terms as "high", "medium" and "low" quality. The Assessment component provides an evaluation of the pragmatic quality of the overall information and enables the communication of where problem areas might exist through an expression of the (un)reliability associated with the quantitative information conveyed in the Numeral, Unit, and Spread categories. Finally, _Pedigree_ is the most qualitative and complex of all categories that provides an overall quality measure by communicating the scientific basis underlying the methods that were used to actually produce the quantitative estimates. This basis conveys the comparative border with ignorance by displaying what more powerful means were _not_ deployed in the production of the information (Funtowicz & Ravetz 1990). Together Assessment and Pedigree express conceptual operations of criticisms and evaluation on the entries in numeral, unit, and spread categories. Therefore, they tend to be very reflective in nature and can foster a degree of quality "self-awareness" among users.

NUSAP provides the first framework for explicitly incorporating the issues of information quality into the communication of scientific estimates. While the different components of NUSAP have been individually considered previously in scientific documentation (included generally through accompanying text, footnotes and appendices), these earlier expressions of information quality tended to provide an ineffective mechanism for communicating the impact from these components and were, therefore, often ignored in policy formulation. The NUSAP formalism places the qualitative components into as prominent a position as the specific numerical estimates, thereby communicating both the quantitative estimates and the quality of these estimates concurrently. Thus, NUSAP can be used to indicate important trade-offs between uncertainty and reliability and quality. For instance, while decreasing the precision of an estimate might increase its reliability, the actual quality of the data may not have been sufficient to justify the higher precision in the first place.

Although NUSAP provides a general methodology for communicating both information and uncertainty, it provides no explicit guidelines for actually developing the specific qualitative categories related to uncertainty and reliability (i.e. the Assessment and Pedigree categories). Furthermore, the subjective operations related to the assessment and pedigree categories can neither be performed by "automatic" means nor can they be accomplished in isolation from the whole body of relevant scientific knowledge. Hence, in order to communicate effectively using NUSAP, the various stakeholders would be required to become familiar with and to develop extensive craft skills in NUSAP methodologies, in addition to those skills required for the specific application under study. Since many of NUSAP's craft skills would be of greater familiarity to scientists than to the laypublic (i.e. non-scientists), this would necessitate that sufficiently detailed explanations would need to accompany any application of NUSAP in order for it to be effective in an EA. It would appear that the convolution provided by these supplementary technical-skill requirements serve only to add unnecessary extra layers of complexity to the communication of the climate change uncertainties.

5.3.3 WRI Qualitative Narratives

The World Resources Institute (WRI) presented a report that sought to identify gaps in data and information on global ecosystems in which the data quality was described using qualitative narratives and data descriptors such as *lacking*, *not reported*, *not available*, *good-quality*, *anecdotal*, and *expert opinion* (WRI 2000). An example of a narrative statement illustrating a lack of information might be that the "data are limited by a difficulty in identifying species and assessing their impact" (Inch 2001). While this narrative approach might provide the only possible way to communicate the extent of situations covered in an analysis devoid of sufficient quantitative data (such as that undertaken by the WRI), a broader applicability of the method is effectively hindered by its inability to provide any sound basis for comparison across different issues. Since such a comparison would be requisite in almost all EAs, the use of such general narratives to communicate different climate change uncertainties is rendered somewhat ineffectual. Furthermore, climate change studies can produce an abundance of quantitative information (as demonstrated in the previous chapter) and can therefore consider more extensive methods for impact communication.

5.3.4 EPA Qualitative Method

An analysis by a Commission of the Environmental Protection Agency (EPA) into methods that could be used to reflect uncertainties in measurement and estimation techniques concluded that qualitative descriptions were needed for most risk assessments, but determined that a quantitative uncertainty analysis of risk estimates was seldom necessary (EPA 1997a, 1997b; Inch 2001). The EPA report placed considerable emphasis on distinguishing variance from uncertainty, where variance was considered to be the natural diversity/variability that could be known, while uncertainty represented unknown or only partially known information. While the Commission strongly supported the use of mathematical descriptions of variability, it remained very "…doubtful that much value would be added…by formal mathematical analyses of uncertainty". However, it was recognized that words used as descriptors could hold very different meanings to different stakeholders and therefore these descriptors should be *calibrated* according to a quantification translation (similar to that in the IPCC study above). Unfortunately, as with the IPCC recommendations, the details required for such quantification prove insufficient to characterize the multiple climate change uncertainties inherent within EAs and such a deficiency therefore renders such an approach inadequate for direct application to EAs.

5.3.5 RIVM Uncertainty Frameworks

The National Institute of Public Health and the Environment in the Netherlands (acronym RIVM) created a "universally encompassing" taxonomy of all types of uncertainty from any subject area by identifying the two meta-level sources of uncertainty; *variability* and *limited knowledge* (van Asselt *et al.* 2001). Under this scheme, the sources of variability uncertainty were categorized into: the inherent randomness of nature; the value diversity among people; the unpredictable, macro-level societal processes of social, economic and cultural dynamics; non-rational human behaviour; and, unexpected developments and/or consequences arising from technological surprises. The sources of limited knowledge uncertainty were categorized into: inexactness arising from lack of precision, metrical uncertainty and/or measurement error; a lack of observations and measurements; things that are immeasurable in practice; conflicting evidence; processes currently unknown, but likely to become known; processes that are understood in principle, but can never be fully predicted; and, processes that cannot be determined. Although many of these sources of uncertainty are valid, not all are particularly useful (i.e. processes that cannot be determined) for decision making (Inch 2001).

RIVM developed a comprehensive framework for structuring an assessment of these uncertainties that combined a technical analysis with processes derived from social theory (van Asselt *et al.* 2001). The five primary steps in this framework involved: (i) identifying the starting perspectives and different worldviews; (ii) identifying and categorizing the relevant uncertainties and ranges for the relevant worldviews determined in (i); (iii) using the relevant worldview scenarios to generate future scenarios; (iv) comparing the future scenarios in terms of risk; and, (v) performing a quality review of the process. A full application of the framework necessitates that lengthy consultations be undertaken to identify the perspectives of the various stakeholders. When multiple perspectives exist (as would be the case in almost any EA), the magnitude of work required to conduct all of the steps in the framework can quickly prove intractable. While this framework is valuable because it integrates technical and value-based human aspects into complex system principles, it does not propose any of the tools requisite for performing risk comparisons. Since this approach only supplies a general framework, it provides only a limited tool for assessing uncertainty. Hence, its applicability (if any) for communicating climate change uncertainties in EAs would be very restricted.

5.3.6 Stirling Frameworks

Similar to the RIVM approach, Stirling (1998) produced a structured framework method for categorizing uncertainties, but with a guidance as to methods that could be used to address the uncertainty type. This framework is summarized in the following table (Stirling 1998).

Table 5.3: **Stirling Framework for Categorizing Uncertainties**

	OUTCOMES	
LIKELIHOODS	**Known**	**Poorly Defined**
Firm basis for probability	**Risk** Apply Frequentist Methods	**Ambiguity**
Weak basis for probability	**Risk** Apply Bayesian Methods	Apply Sensitivity Analysis
Unknown	**Uncertainty** Apply Scenario Analysis	**Ignorance** Apply Precaution

The Stirling framework can be applied to dynamic processes where the changes occur on a temporal basis and characterizes situations into either risk, ambiguity, uncertainty, or ignorance. These framework distinctions can be more broadly communicated as referring to those types of situations: (i) with known outcomes and estimated probabilities; (ii) with known outcomes but with unknown likelihoods; (iii) with unknown outcomes and estimated probabilities, or (iv) with unknown outcomes and with unknown likelihoods. The Stirling framework can be considered robust in content but weak as a communications device, since it contains terms that are both easily confused and expressed in language terms that have "common usage" connotations that are quite different from the specific intent of the framework (Inch 2001). Hence, in order to actually apply the Stirling framework, the target audience would need to both learn the classification system and apply new definitions to commonly used words. Such user-learning requirements could prove problematic in EAs given the wide diversity of stakeholder skill-sets.

5.3.7 Meta-Level Qualitative Methods

Richards & William (1999) introduced a "meta level" descriptive approach that partitioned uncertainties into four fundamental classification schemes. These schemes described the nature of uncertainty as being either: (1) *temporal* based upon both future and past states; (2) *structural* due to its complexity; (3) *metrical* in its measurement applications, or; (4) *translational* in explaining uncertain results. Inch (2001) describes how this scheme classification would be very difficult to apply to any complex system that has not already achieved steady-state. Given that the impacts from climate change are expected to occur on an escalating basis over the course of several centuries and that a steady-state could only occur if GHG emissions are effectively controlled (and all feedbacks are inconsequential, see Chapter 2), it would appear that the requisite conditions for the application of this method would not be effectively satisfied. Hence, such a meta-level, qualitative style of analysis does not seem to hold any direct promise for effectively addressing how the assessment and communication of climate change uncertainties could be incorporated into EAs.

5.4 Communicating Climate Change Uncertainties in EAs

Under specific circumstances, each approach reviewed in the previous section possesses beneficial features that could prove useful for communicating the uncertainties of climate change to the disparate stakeholders in project EAs. While the methods provide numerous alternatives for uncertainty communication, there is an underlying requirement for stakeholders to have acquired significant craft-skill proficiency in order to understand each approach and this obligation for supplementary technical skill possession introduces the major hurdle to their implementation. Any requirement for additional craft-skill acquisition by all stakeholders must be considered a negative feature sufficient to preclude the method's widespread adoption. Consequently, the overarching observation must be that none of these individual methods supplies a universally adaptable framework that could become the "gold standard" for communication.

While global climate change is anticipated to occur in the future, the timing and magnitude of the resulting impacts contain numerous uncertainties and it is unlikely that these uncertainties can be reduced in the near-term. In spite of these uncertainties, mitigation and adaptation steps must be taken in projects to address the range of possible outcomes and, when appropriate, various contingencies need to be included in the projects. Given the needs of decision-makers to weigh potential responses to climate change risks before all of the uncertainties can be resolved, the available information (imperfect as it may be) must be synthesized, evaluated and communicated in a responsible and informative manner (Moss & Schneider 2000; Ravetz 1986). Furthermore, for projects that could be contributing to climate change through the production of greenhouse gases, there exist legislated requirements that such contributions be explicitly addressed within an EA (the Canadian Environmental Assessment Act, Sections 2 & 16).

Although the goal of a comprehensive communication framework remains an elusive concept, it is imperative that the potential impacts from the uncertainties be effectively communicated to the different stakeholders in EAs, irrespective of the intrinsically obvious difficulties in this communication. With the considerable uncertainties and disagreements about climate change, this communication of the uncertainties must accomplish two fundamental tasks: (1) Clearly communicate the results of the analyses, including the range of impacts that might be caused by climate change; and, (2) Clearly communicate information about the degrees of belief in, and acceptance of, these results. Recognizing that the stakeholders will always be drawn from a wide gamut of constituencies possessing dissimilar technical skill-sets, it is therefore suggested that the following recommendations (that have integrated ideas from the methods in the previous section) be adopted as the minimally prudent level of communication within EAs.

In order to accommodate the disparate craft-skills of the stakeholders, the presentation of information should appear in the form of a comprehensive verbal description written in an accessible, non-technical, clear, and concise format. Since such explanations would necessarily require a significant amount of work on the part of the proponent, detailed verbal descriptions are warranted only for the most significant elements and for the communication of their resulting uncertainties (Moss & Schneider 2000). Hence, it is important for the proponent to identify the *key components* in the EA and to articulate the nature of the major uncertainties inherent within each of these components. Specifically, the proponent, with input from stakeholders, must determine which models, data sets, assumptions, and results constitute the key component elements of the EA process. In order for the stakeholders to be able to assess their degrees of beliefs and confidence in both the resulting estimates and underlying uncertainties, verbal explanations need to be provided concerning the acceptability of each of the key components identified for the EA, namely: (i) The *models* used, (ii) the *data* sets employed, (iii) the major *assumptions* made, and (iv) the *results* achieved. Providing information about these factors can prove to be challenging, but the inclusion of clear and detailed summaries of each key component is particularly important in the verbal documentation. Sufficient care must enter into this process, to ensure that any technical meanings behind data presented in the description makes the underlying implication readily accessible to the layperson.

Hence, the following interrelated types of information should be provided within the verbal description to help assess the degrees of acceptability of each identified key component. For each *model* that was used: (i) the source(s) should be identified; (ii) the degree to which it is an accurate representation of reality should be stated; (iii) whether it is based on an established underlying theory or school of thought; (iv) whether the model has undergone peer review; and, (v) degree of acceptance of the model by the overall research community. For each set of *data* employed: (i) the data source(s) need to be identified; (ii) whether it is primary or modified (secondary) data; and, (iii) whether or not the data is based on an established theory or school of thought. For each key *assumption* made: (i) the degree to which it is an accurate representation of reality should be stated; and, (ii) the degree of acceptance of the assumption by the research community should be ascertained. For each set of resulting *estimates*: (i) it should be stated whether these have been independently reviewed; and, (ii) the degree of acceptability by the reviewers needs to be stated. A summary assessment regarding the general level of *overall confidence* in the component would also be required.

The scenarios used as the data source in the previous chapter's hydro example can be considered as one of that study's key components. If a verbal assessment were applied to these scenarios, the resultant description might proceed in the following way. The identified <u>sources</u> for the scenario *models* used in the example were the IPCC and the CICS. These scenarios could be thought of as reflecting a particular <u>school of thought</u> as to future climate change outcomes, yet their actual representation of reality at this particular point in time is <u>unknown</u>. The scenarios have been subjected to extensive <u>peer review</u> due to ongoing worldwide use, but the general acceptance of them can only be considered <u>variable</u>, since there exist numerous climate change "dissenters"–thus, their degree of acceptance depends inherently upon the specific viewpoints held by the user. There are <u>various</u> *sources* for the data in the scenarios, but overall the figures provided should be considered as <u>primary</u> data. The *key assumptions* used in the scenarios can be considered as only a <u>medium</u> representation of reality, since the scenarios have been constructed from consensus opinions (i.e. highly "extreme" beliefs will have been averaged out) and, as such, the acceptance of these assumptions can be quite <u>variable</u>. The *resulting estimates* of the scenarios <u>have been</u> independently reviewed and their overall acceptance can be considered as being accepted at the <u>medium</u> level–implying that while many people currently accept these scenarios as the best expert judgment available at this time juncture, there exist numerous dissenters and/ or supporters of alternative interpretations. Therefore, the *overall confidence* in the scenarios used in the hydro example can be considered as being only at a <u>medium</u> level of acceptance.

In addition to describing the uncertainties in the key components, the relevant documentation will necessarily have to incorporate and present both quantitative and qualitative summary information–also expressed in an accessible verbal fashion. In order to convey the uncertainties in the *quantitative* information, the following types of summary information should be described within the body of the narrative (with the possible inclusion of supporting figures and tables): (i) mean values and variances of the estimates; (ii) confidence intervals of the estimates where possible; (iii) the ranges of the estimated values with a noting being made of the possible extreme values, in particular; (iv) full probability distributions of the estimated impacts; (v) detailed descriptions of important thresholds and vulnerabilities that have been uncovered in the study; and, (vi) descriptions of any other significant quantitative estimates determined in the analysis.

For impacts that are measured qualitatively, the uncertainties can be described and presented only with considerably less precision. To support the inclusion of *qualitative* information, the following types of summary descriptions should be provided for the uncertainties (again with possible supporting tables and ancillary devices): (i) descriptions of the central tendency of the impacts together with any possible variation away from it, such as "most likely to have a medium impact with a significant possibility of attaining a high level"; (ii) the ranges of the estimate, perhaps described as "low to medium" ; (iii) an explanation of any important thresholds and vulnerabilities such as "a possibly significant loss of a specified game species"; and, (vi) descriptions of any other significant qualitative impacts determined in the analysis. Furthermore, whenever imprecise, qualitative terms and descriptors such as "low", "high", or "significant" have been used, the basis underlying their particular application needs to be clearly articulated.

While verbal descriptions should be considered as the minimum required standard for communicating climate change uncertainties in EAs, their use need not preclude the additional application of other methods and/or frameworks (i.e. like those from the previous section)–should the proponent believe that these methods provide ancillary support to the overall presentation. These additional techniques are appropriate only when they can contribute broadly understandable additional support to the verbal descriptions without the need for craft-skill acquisition beyond the realm of the stakeholders. Thus, the proponent should be aware that communicating uncertainties to a diverse set of stakeholders using supplementary techniques might also require the parallel addition of extremely detailed verbal descriptions in order for their actual meaning to remain accessible to the non-technical stakeholders.

An example of one evaluative framework applied to three key interrelated components (the scenarios used, the streamflow model used, and the energy model used) of the previous chapter's hydro example appears in the table below. Each column in the table corresponds to a summary evaluation of the specific component. While proponents would need to carefully explain the purpose and contributions of such an ancillary table, the primary descriptive vehicle would remain the verbal descriptions of the uncertainties in the components and the related issues. Thus, it should be noted that the column evaluating the scenarios actually provides a summary of the verbal description that appeared earlier in this section. Similar verbal descriptions would also be required for the streamflow and energy models.

Table 5.4: **Presentation on Acceptability: Scenario Analysis for Energy**

		Stage 1: Scenarios	Stage 2: Streamflow	Results: Energy
Model:	Source	IPCC/CICS	Consultant	Consultant
	Rep. of reality	Unknown	Medium	High
	Theory/Sch.of thought	School	Est. Theory	Est. Theory
	Peer review	Yes	No	No
	Acceptance	Variable	-	-
Data:	Source	Various	MNR	-
	Primary/Sec.	Primary	Primary	-
	Theory/Sch.of thought	-	-	-
Key Assumptions:				
	Rep. of reality	Medium	High	High
	Acceptance	Variable	High	High
Resulting Estimates:				
	Indep. review	Yes	No	No
	Acceptance by review	Medium	-	-
	Overall confidence	Medium	Low-Medium	Low-Medium

Additionally, whenever estimated impacts result from the product of a series or concatenation of components, the overall quality and acceptability of these results must necessarily depend on the combination of the quality of each of the individual models, data and assumptions involved. In order to communicate the acceptability of these combined results, an evaluation of the acceptability of each key individual constituent component would be requisite, together with an *overall assessment* of their combination.

5.5 Conclusion

When a consistency of purpose exists among the users of information, then a single superior approach may exist that can universally articulate the uncertainties in a process to all concerned parties. This chapter has reviewed a number of the possible methods that could be considered for communication of the climate change uncertainties to stakeholders in project EAs. Some of these methods would be appropriate for application to specific types of project EAs, but not for others due to the requirements for significant supplementary craft-skills amongst the stakeholders. Clearly, proponents should decide which methods are applicable to their EA and these methods should be used wherever they are deemed appropriate. However, since an EA process necessarily includes an extremely broad spectrum of stakeholders possessing widely divergent craft-skills, perspectives and viewpoints, the existence of any single universally acceptable method for communicating climate change uncertainties is most unlikely.

Notwithstanding the lack of a single method for communication, the issues of climate change must be considered in project EAs and the resultant uncertainties must be communicated to stakeholders. The major suggestion resulting from this chapter is that these EAs should communicate the uncertainties using a "verbal" description of the climate change uncertainties, ensuring to include, at a minimum, certain specific information within this narrative approach. The goal of the verbal description is to circumvent the craft-skill deficiency difficulties inherent in the EA's stakeholders by clearly communicating the major uncertainties behind all of the key components in a format that is accessible to all stakeholders. Additional, ancillary methods, such as graphical representations and/or summary frameworks, can also be employed wherever such inclusion facilitates and supports the communication of the uncertainties with even greater clarity. However, proponents should incorporate these additional methods with the recognition that their inclusion may not be well-understood by all stakeholders and a proviso within the EA should indicate that these methods support the more verbal descriptions, but are not replacements for them.

6

Recommended Guidelines and Conclusions

6.1 Introduction

While numerous observers have recognized the potentially catastrophic conse-quences associated with a changing climate, several studies have indicated that the impacts and uncertainties of climate change have been inadequately addressed in project environmental assessments (EAs). Since project decision-making can be adapted in many ways to respond to climate change issues, project proponents are obliged to consider potential climate change impacts and uncertainties in their analysis.

This study has examined how various uncertainties from climate change can be analyzed, incorporated, and communicated within project EAs. Obvi-ously proponents would need to modify the approaches presented to the situa-tions faced within their specific EAs in order to include the relevant climate uncertainties into their own specific analyses. This final chapter concludes with a synopsis of the various suggestions and recommendations for accom-plishing these tasks that have been discussed throughout the study.

6.2 Recommended Guidelines for Incorporating Climate Change Impacts and Uncertainties into an EA

In practice, most projects have been designed on the basis of historical conditions, but such historical specifications will not remain applicable under many of the conditions of a changing climate. Since there are numerous ways in which climate change impacts could be adapted to at the project level, these approaches and the extent of the possible impacts should be addressed during the project's EA. Consequently, there is a need in EAs for proponents to consider the potential impacts and uncertainties from climate change. However, the major difficulty for proponents is to determine both how changing climatic conditions could affect the project and how the project could affect climate change, and also how to effectively incorporate the resulting uncertainties into the EA. While there can be no definitive rule for selecting one approach for addressing climate change uncertainties over another, this section provides suggestions and observations that should be considered during the EA process.

1) *The environmental effects related to climate change that should be examined at the project EA level need to be considered from three categorical perspectives:*

   (i) the effects of climate change on the project;

   (ii) the effects of climate change on the impacts resulting from the project;

   (iii) the effect of the project on GHG emissions.

Direct climate change impacts on a project should be considered, since these are exactly the types of "environmental effects" that are studied in any project EA. Changes that a project may cause within the environment should be examined in relation to the state of the environment without the project and, since such changes may be affected by climate change, should be duly considered in its EA. Since a project directly contributes to climate change by either the production or the reduction of GHG emissions, estimates of the levels of these emission contributions/reductions needs to be made in the EA.

2) *To estimate a project's effective contribution to climate change, its anticipated production of–or reductions to–greenhouse gas (GHG) emissions should be disclosed relative to its specific industry sector target.*

While acknowledging that an individual project's contributions to global climate change may appear infinitesimal when viewed from a global perspective, each project's implications for climate change should be addressed in its EA. One reasonable surrogate estimate for these potentially wide-ranging impacts can be captured by the project's anticipated GHG emissions. By contrasting these GHG emissions relative to broad national industry sector emission reduction targets (for instance, the industry sector reduction targets specified in Canada (2002)), an effective proxy measure for the overall climate change impacts from the project can be provided. However, a complete life-cycle analysis should also be conducted in order to determine the full impact of the project on GHG emissions.

3) *For a project EA, three major analytical methods that should be considered to evaluate the impacts and to incorporate the related uncertainties of climate change are:*

(i) scenario analysis;

(ii) probabilistic analysis;

(iii) sensitivity analysis.

While scenario analysis is the analytical method most often associated with climate change studies, sensitivity analysis and probabilistic analysis represent more general techniques that have been more widely used in addressing uncertainties. Depending upon the particular EA, each of the three methods, or various sequences and combinations thereof, could be deemed most appropriate for conducting a specific analysis. Proponents would need to be able to justify the specific choice of analytical method(s) to employ.

4) *Key factors in the selection of an appropriate analytical method to evaluate the impacts and to incorporate the related uncertainties of climate change should include:*

(i) the measurability of the data;

(ii) the level of difficulty in using the method;

(iii) the level of modelling effort required.

A key factor in the choice of the specific analytical method to be used is based upon the nature of the measurability of the data. Hence, the applicability of a method can depend upon whether the values it requires as inputs are well-defined and quantitatively measurable or are only ill-defined and qualitatively measurable. The level of difficulty in the use of a method plays a major role in the appropriateness of its selection. Methods requiring significant use of resources (i.e. expertise, time, data, cost, computing) should not be used to study less important impacts. Methods requiring significant modelling effort necessitate the existence of a clearly developed level of understanding about the relationships between climate change and the specific impact (i.e. the existence of well-developed analytical models).

5) *Two major factors that should influence the choice of the analytical approach for addressing climate change in an EA are:*

(i) the importance to the project of the specific impact being studied and the importance of the information resulting from the analysis;

(ii) the quality of the models that are available to study the impact and the quality of the quantitative data that is available for use in the models.

6) *Scenario analysis should be the most appropriate approach when the quality of the model is high and quantitative data availability is substantial, and when the impact being studied is of significant importance to the EA.*

Scenario analyses require more extensive computational efforts, better developed models, and a better quality of quantitative data than sensitivity analyses. Hence, scenario analyses prove to be the most appropriate choice when the quality of the model is high, the quantitative data availability is extensive, and when the impact being studied is of major importance. These requirements will often be satisfied for impacts directly linked to climate change variables because of the extensive development of climate change scenarios that has already been performed by numerous international scientific institutes.

7) *The first step in scenario selection is for proponents to identify exactly which scenarios meet the data requirements for the variables needed in their analysis.*

The website for the Canadian Climate Impacts Scenarios (CCIS) permits proponents to download extensive scenario data of the possible future climate produced by various major international climate institutes. However, several of these scenarios either possess restricted sets of data or do not contain variable estimates covering all applicable future time periods. Therefore, proponents would necessarily be restricted to selecting data from those scenarios containing the information for all of the required climate variables over the time horizon of their analysis. Fortunately, summary information on the data availability of each scenario is provided by the CCIS.

8) *Due to different scaling resolutions, proponents should ensure that the weather data taken from different scenarios contains the specific region in which their project actually occurs.*

The size of specific regional grid references of the different climate institutes can cover dissimilar geographic regions due to the various non-uniform scale resolutions employed in scenario construction. Details on the different scale resolutions of the scenarios can be found on the CCIS website.

9) *Should there be insufficient time to use all scenarios meeting specified data requirements, then it is important to select <u>enough</u> scenarios to bound the full range of scenario results for the relevant climate variables.*

Since all of the climate change scenarios could occur, it is imperative for a proponent to consider a full spectrum of design options needed to account for the uncertainties in the decision process—including a due consideration for any "extreme" events. If circumstances prevent an examination of the project operating under all available scenario options, then proponents need to select specific scenarios that represent the extreme ranges of the key variables required in the analysis, as well as more moderate, intermediate scenarios. The CCIS website provides several useful suggestions for such restricted scenario selection.

10) *In any study influenced by climate change impacts, it is essential to consider a range of different scenarios so that the distribution of possible outcomes can provide a useful context for understanding the relative likelihoods of various occurrences.*

Once a proponent has determined which scenarios possess the needed variables over the required time periods, a decision must be made as to which scenarios to include in their EA. The scenarios should be selected in a fashion consistent with international methodologies. Namely, proponents should apply multiple scenarios that span a range of possible future climates and should include scenarios constructed by at least two different climate institutes. Reviewing as many scenarios as possible provides a broader context of what is likely to happen and how the key variables might change in the future.

11) *Scenario analysis can produce an abundance of quantitative and graphical data that provides proponents with considerable information on the possible reliabilities and uncertainties that could be encountered by a project under alternate climate change futures.*

Scenario analysis can be used to produce the representative range of climate change futures recommended by the CICS to support the analysis in EAs. The information produced by scenarios can be used to support the planning implications from using various different decision criteria for project design (such as: best-case-worst-case analysis, decisions based on averages or ranges of values), while simultaneously being used to answer questions on the environmental implications arising from such decisions.

12) *Scenario analyses can be used in numerous ways to address different decisions and the information produced should provide data on the ranges, expectations, and uncertainties inherent in the project in the face of a changing climate. The values produced by evaluating a project design based upon the conditions of one scenario while operating under the conditions of several alternate scenarios could be used to provide estimates of the uncertainty in the performance of the project over the time horizon.*

It can be of considerable interest to explore the consequences of making decisions based upon the use of any one scenario and determining the subsequent impacts should, in fact, any of the remaining scenarios actually occur. If these scenarios could be considered as representative of all possible future climate change paths, or at least their extremes, then the range of values determined in this output would provide the possible extreme. The determination of such limits could prove to be essential information in an EA.

Scenarios can be considered as representations of the best projections of future weather patterns that current scientific knowledge can provide and should, therefore, be considered as plausible representations of the range of probable futures resulting from climate change. Thus, it makes sense to consider these possible futures when making key decisions regarding a project's design. Since no single scenario can be treated as more or less probable than others, the likelihood that any one scenario could occur would be the same as that for any other scenario. Hence, it would be quite reasonable to design a project on the basis of any one particular scenario and to investigate its performance over the entire time horizon under the assumption and conditions that every other scenario had, in fact, actually occurred. Therefore, the range of values and outcomes produced during such an analysis could provide an effective estimate of the uncertainty in the performance of the project over the applicable time horizon.

13) *Proponents should practice vigilance against the adoption of a false sense of precision in finer resolution scenario data, since downscaled data, using some form of interpolation technique, has necessarily been employed in its creation.*

Since any predictions of weather impacts on geographically-localized projects require the use of very detailed climatic information, the weather variables most needed for the majority of project EAs require very specific regional details. However, it is impossible for climate models to provide certainty about specific weather variables in specific locations due to the resolutions actually used in the scenario computations. The CCIS website describes how it is possible to downscale the coarser-scaled scenario data from the various climate institutes into much smaller grid references and does include scenario data expressed at much finer resolutions than that provided by the original scenario analyses. The CCIS provides numerous useful instructions, together with several caveats, on how to effectively employ this finer resolution data in an analysis.

14) *To justify the significant development time, computational efforts, and economic costs needed for a probabilistic analysis, the importance of the impact being studied should be high.*

Probabilistic methods require the existence of both well-developed models and well-defined, quantitative data for input into the models. Probabilistic analysis also requires considerable technical expertise and effort on the part of the user. Therefore, in order to justify the significant development time, computational efforts, and economic costs needed for probabilistic studies, the importance of the impact being studied should be high.

15) *If an impact can only be measured qualitatively or cannot be defined probabilistically, then probabilistic methods should not be employed.*

16) *Sensitivity analysis can be an appropriate method when either the models and data required for the scenario and probabilistic methods are not readily accessible or when the importance of the studied impact relative to the project is low.*

Sensitivity analysis can produce a good first-step in most analyses, since it can essentially be applied as an analytical screening device in virtually all cases. If the impact being studied can be shown to be of relatively minor importance, then additional analysis would not be required. Given its wide applicability, sensitivity analysis can provide the only choice when it is not possible to employ any other analytical methods. Thus, sensitivity analysis can prove to be the appropriate analytical method when either the models and data required for scenario and probabilistic methods are not readily accessible or when the importance of the studied impact to the project is relatively low. Unfortunately, sensitivity analysis cannot be performed on anything more than a very narrow set of parameters due to the combinatorial explosion in its computational requirements.

17) *Proponents should use sensitivity analyses to identify which critical climate variables significantly impact their project and which project design vulnerabilities to focus upon given the inherent uncertainties in predicting the changes to these variables.*

For a project, the proponent might want to determine the sensitivities that prove essential to effective decision-making. To perform this process, the proponent could employ expert judgment to select several key climate variables believed to have significant impact on the project and use sensitivity testing to determine whether a more detailed climate change evaluation was warranted. The proponent should involve both experts and stakeholders in the determination of which climate variables to actually test for sensitivity. If the thresholds of a few key climate variables that subjected the project to major sensitivities could be determined, then a subsequent examination of the vulnerability of the project to the uncertainties in these climate variables would be requisite. Knowledge of project sensitivities could direct additional impact/ sensitivity studies to evaluate what potential design changes could be incorporated in order to adapt to or mitigate against the resulting vulnerabilities. Hence, sensitivity analysis can be used to identify the critical climate variables that could significantly impact the project and which project design vulnerabilities to focus upon given the inherent uncertainties in predicting the changes to these climate variables.

18) *If pathologically extreme values for any climate parameter in a sensitivity analysis demonstrated no significant impact on a project, then this would indicate that the tested parameter need not be evaluated further in the EA.*

In order for sensitivity analyses to become computationally manageable, numerous restrictive assumptions must be adopted for the key climate parameters and no singularly best process exists for specifying the steps in conducting sensitivity studies. The ranges and/or numbers selected for the climate parameters must be sufficient to determine whether their impact is significant in the decision-making process. If a sensitivity analysis used parameter values beyond those ever likely to occur (i.e. values representing pathological extremes) and demonstrated that these extreme values had no significant impact on a project, then this would indicate that the tested parameter was insignificant and need not be evaluated any further.

6.3 Recommended Guidelines for Communicating the Climate Change Uncertainties in an EA

Since stakeholders in EAs receive an open forum for expressing their opinions regarding any proposed management of environmental impacts, it is imperative that the uncertainties surrounding climate change be somehow effectively communicated to *all* of the disparate set of stakeholders. While numerous communication methods exist, a significant impediment to their widespread adoption in EAs is that to be understandable, most of these approaches require the stakeholders to possess sophisticated supplementary technical-skill proficiencies. Therefore, irrespective of the intrinsically obvious difficulties in communication and recognizing that stakeholders will always be drawn from a diverse constituencies possessing dissimilar technical skill-sets, it is recommended that guidelines 19-33 be adopted in project EAs as the minimally prudent level of communication for climate change uncertainties.

19) *The communication of the uncertainties about climate change should accomplish two fundamental tasks:*

   (i) Clearly communicate the results of the analyses, including the range of impacts that might be caused by climate change;

   (ii) Clearly communicate information about the degrees of belief in, and acceptance of, these results.

20) *In order to accommodate the disparate technical-skills of the stakeholders, the presentation of information on climate change uncertainties should appear in the form of a comprehensive non-quantitative description written in an accessible, non-technical, clear, and concise format.*

The issues of climate change should be considered in project EAs and the resultant uncertainties need to be communicated to stakeholders. Since the EA process can include an extremely broad spectrum of stakeholders possessing widely divergent technical-skills, perspectives and viewpoints, the existence of a universally applicable method for communicating climate change uncertainties is most unlikely. The goal of a non-quantitative written description must be to circumvent the technical-skill deficiency difficulties inherent in the stakeholders by clearly communicating the major uncertainties behind all of the key components in a format that is readily accessible to all stakeholders. Additional, ancillary methods can also be employed whenever their inclusion facilitates and supports the communication of the uncertainties with even greater clarity. Proponents should decide which methods are most applicable to their specific EA and these methods should be used when appropriate. However, proponents should only incorporate these additional methods with the recognition that their inclusion may not be well-understood by all stakeholders and a stipulation within the EA should indicate that these methods support the non-quantitative written descriptions, but are not replacements for them.

21) *In addition to the non-quantitative written description, other methods and frameworks for communicating climate change uncertainties should be incorporated into an EA, if the proponent clearly believes that these methods provide ancillary support to the overall presentation of the analysis.*

While non-quantitative written descriptions should be considered as the primary descriptive vehicle and the minimum required standard for communicating climate change uncertainties in EAs, their use need not preclude the additional application of other methods and/or frameworks. Several of these types of communication methods and frameworks are described in Chapter 5. However, these additional techniques are appropriate only when they can contribute broadly understandable additional support to the non-quantitative written descriptions without the need for technical-skill acquisition beyond the realm of the "common" stakeholder. Thus, proponents should be aware that communicating uncertainties to a diverse set of stakeholders using supplementary techniques might also require the concurrent addition of extremely detailed non-quantitative written descriptions in order for their actual meaning to remain accessible to the non-technical stakeholders.

22) *If proponents employ methods for communicating information regarding the impacts of the climate uncertainties involved in their EAs beyond the non-quantitative written description, then they should carefully assess the technical sophistication level of the stakeholders in order to determine the appropriate means for communicating the information.*

Proponents would need to determine whether the non-quantitative written description sufficiently captured and communicated the nature of the changes and impacts uncovered in their analysis or whether other presentation methods, or combinations thereof, might be more helpful to support an effective analysis. Alternate approaches to non-quantitative written descriptions might exist that could communicate the information to the EA's stakeholders more effectively.

Histograms and other similar graphical representations can provide an accessible visual representation of uncertainty in output data. Although not as visually appealing, tabular formats permit explicit side-by-side comparisons of outcome likelihoods under different assumptions of climate change and could include additional columns of historical likelihoods for comparative purposes. While tabular representations could prove difficult for stakeholders to comprehend, the actual information contained within them can convey considerably more analytical details than more visually accessible figures.

The approach(es) actually selected by the proponent would necessarily have to depend upon the complexity of the output, the type of information that had to be communicated, and the technical sophistication level of the interested stakeholders. Hence, proponents need to carefully determine the level of technical sophistication of the stakeholders in order to assess the appropriate means for communicating the climate change uncertainties discovered in their analysis.

23) *Proponents may necessarily need to communicate the results of a probabilistic analysis using a combination of several different methods.*

A probabilistic analysis provides challenges as to the appropriate way to convey the meaning of stochastic impacts, since simulation outputs tend to be expressed in the form of probability distributions rather than single numerical values. Sometimes the impacts can be best expressed as visual figures of the probability distributions, while at other times they might be better expressed numerically as average values or ranges. Hence, proponents may necessarily need to communicate their probabilistic analyses using a combination of more than one method.

24) *Proponents should identify the <u>key components</u> in an EA and should articulate the nature of the major uncertainties inherent within each of these key components. Proponents need to provide non-quantitative written explanations concerning the acceptability of each of the key components identified for the EA. Namely explanations of:*

(i) the <u>models</u> used;

(ii) the <u>data sets</u> employed;

(iii) the major <u>assumptions</u> made;

(iv) the <u>results</u> achieved.

Detailed non-quantitative written descriptions are warranted only for the most significant elements and for the communication of their resulting uncertainties. Proponents should determine and communicate which models, data sets, assumptions, and results constitute the most important constituent elements, or *key components,* of the EA and the major uncertainties inherent within each of these key components. Hence, it is important for proponents to identify the key components in the EA and to articulate the nature of the major analytical uncertainties inherent within each of these components.

In order for the stakeholders to be able to assess their degrees of belief and confidence in both the resulting estimates of climate change impacts and their underlying uncertainties, non-quantitative written explanations should be provided concerning the acceptability of each of the key components identified for the EA. These are: (i) the *models* used, (ii) the *data* sets employed, (iii) the major *assumptions* made, and (iv) the *results* achieved. Providing information about these factors will prove to be challenging, but the inclusion of clear and detailed summaries of each key component is particularly important in the non-quantitative written documentation. Sufficient care should enter into this process, to ensure that any technical meanings behind data presented in the description makes the underlying implication readily accessible to the layperson.

25) *For each _model_ that was used, the following interrelated information should be provided within the non-quantitative written description to help assess its degrees of acceptability:*

(i) its source should be identified;

(ii) the degree to which it is an accurate representation of reality should be stated;

(iii) whether it is based on an established underlying theory or school of thought;

(iv) whether the model has undergone peer review;

(v) the degree of acceptance of the model by the overall research community.

26) *For each set of _data_ employed, the following interrelated information should be provided within the non-quantitative written description to help assess its degrees of acceptability:*

(i) the data source needs to be identified;

(ii) whether it is primary or modified (secondary, tertiary, etc.) data;

(iii) whether or not the data is based on an established theory or school of thought.

27) *For each key _assumption_ made, the following interrelated information should be provided within the non-quantitative written description to help assess its degrees of acceptability:*

(i) the degree to which it is an accurate representation of reality should be stated;

(ii) the degree of acceptance of the assumption by the research community should be ascertained.

28) *For each set of resulting* <u>estimates</u>, *the following interrelated information should be provided within the non-quantitative written description to help assess its degrees of acceptability:*

> *(i) it should be stated whether these have been independently reviewed;*

> *(ii) the degree of acceptability by the reviewers needs to be stated.*

29) *A summary assessment regarding the general level of* <u>overall confidence</u> *in each of the key components would also be requisite.*

30) *In order to convey uncertainties in* <u>quantitative</u> *information, where appropriate, the following types of summary information, expressed in an accessible non-quantitative written fashion, should be described within the body of the narrative (with the possible inclusion of supporting figures and tables):*

> *(i) mean values and variances of the estimates;*

> *(ii) confidence intervals of the estimates wherever possible;*

> *(iii) the ranges of the estimated values with an explicit note being made of the possible extreme values, in particular;*

> *(iv) full probability distributions of the estimated impacts;*

> *(v) detailed descriptions of important thresholds and vulnerabilities that have been uncovered in the study;*

> *(vi) descriptions of any other significant quantitative estimates determined in the analysis.*

31) *To support the inclusion of* <u>qualitative</u> *information, the following types of summary descriptions, expressed in an accessible non-quantitative written fashion, should be provided for the uncertainties (with possible supporting tables and/or other ancillary devices):*

> *(i) descriptions of the central tendency of the impacts together with any possible variation away from it, such as "most likely to have a medium impact with a significant possibility of attaining a high level";*

> *(ii) the ranges of the estimate, perhaps described as "low to medium";*

> *(iii) an explanation of any important thresholds and vulnerabilities such as "a possibly significant loss of a specified game species";*

(vi) descriptions of any other significant qualitative impacts determined in the analysis.

32) *Whenever imprecise, qualitative terms and descriptors such as "low", "high", or "significant" have been used, the basis underlying their particular application should be clearly articulated in the non-quantitative written description.*

For impacts that have been measured qualitatively, the uncertainties can only be described and presented with considerably less precision than occurs for the quantitative cases.

33) *When estimated impacts result from a concatenation of several different components, proponents should provide an evaluation of the acceptability of each key individual component together with an* <u>*overall assessment of their combination*</u>.

Whenever estimated impacts result from the product of a series or concatenation of components, the overall quality and acceptability of these results must necessarily depend on the combination of the quality of each of the individual models, data and assumptions involved. In order to communicate the acceptability of these combined results, an evaluation of the acceptability of each key individual constituent component would be requisite, together with an *overall assessment* of their combination.

6.4 Conclusion

The phenomenon of climate change has escalated into a major environmental concern of regional, national, and international scope. The major sources of climate change uncertainty revolve around the extent of damage that the earth's climate is already committed to as the climate "catches up" to the greenhouse gas (GHG) build-up that has already occurred and the additional damage associated with future increases in GHG concentrations. Uncertainty has often been cited as a major reason for delaying many of the actions that might mitigate potential impacts from a changing climate, but this position contravenes established principles of sound risk management. However, since the impacts of climate change are both global in scope and essentially irreversible, none of the inherent uncertainties can override the fact that GHG emissions will need to be vigorously and rapidly curtailed.

Clearly, climate change can have significant implications for the environment and for projects that affect the environment and should, therefore, receive due consideration in EAs. Recent reviews of project EAs that have been conducted in Canada have concluded that climate change impacts have been inadequately addressed and that the corresponding climate change uncertainties have been addressed even more poorly. Recognizing these deficiencies, this study has investigated methods for addressing, incorporating, and communicating the inherent uncertainties that surround the various climate change issues into project EAs. Climate change requires immediate attention because of the uncertainties, not in spite of them, and a proper consideration of the uncertainties in EAs increases, rather than decreases, the rationale for preventative action.

References

Ausebel, J.H., 1991, "A Second Look at the Impacts of Climate Change", **American Scientist**, 79, p210-221

Botkin, D.B., J.G. Janak, and J.R. Wallis, 1972, "Some Ecological Consequences of a Computer Model of Forest Growth", **Journal of Ecology**, 60, p 849-872

Campbell, J.M. Jr., J.M. Sr., Campbell, and R.A. Campbell, 2001, Analyzing and Managing Risk, Campbell Publishing, Norman, OK, USA

Canada, 2002, "Climate Change Plan for Canada", Government of Canada, www.climatechange.gc.ca

CCIS 2003, Canadian Climate Impacts Scenarios, http://www.cics.uvic.ca/scenarios/index.cgi

Chociolko, C., 1995, "The Experts Disagree: A Simple Matter of Facts Versus Values", **Alternatives**, 21, 3, p19-25

Clemen, R.T., and T. Reilly, 2001, Making Hard Decisions, Duxbury, Pacific Grove, CA, USA

Colombo, A.F., D. Etkin, and B.W. Karney, 1999, "Climate Variability and the Frequency of Extreme Temperature Events for Nine Sites Across Canada: Implications for Power Usage", **Journal of Climate**, 12, p2490-2502

Dawes, R.M., 1988, Rational Choice in an Uncertain World, Harcourt Brace Jovanovich, San Diego, CA

Decisioneering, 2003, "Crystal Ball 2000", www.decisioneering.com

EPA, 1997a, <u>Mandate of the Commission on Risk Assessment and Risk Management</u>, United States Environmental Protection Agency Advisory Committee Charter, Washington DC, <u>http://www.epa.gov/NCEA/ pres com/riskcom/nr6aa028.htm</u>

EPA, 1997b, <u>Risk Assessment and Risk Management in Regulatory Decision-Making</u>, Presidential/Congressional Commission on Risk Assessment and Risk Management, Final Report, Volume 2, Washington DC, p89

Epstein, P.R., 2000, "Is Global Warming Harmful to Health", **Scientific American**, 283, 2, p50-57

Evans, M., N. Hastings and B. Peacock, 1993, <u>Statistical Distributions</u> (2nd edn). John Wiley & Sons, New York, NY

Funtowicz, S.O. and J.R. Ravetz, 1984, "Uncertainties and Ignorance in Policy Analysis", **Risk Analysis**, 4, 3, p219-220

Funtowicz, S.O. and J.R. Ravetz, 1990, <u>Uncertainty and Quality in Science for Policy</u>, Kluwer Academic Publishers, Dordrecht, The Netherlands

Funtowicz, S.O. and J.R. Ravetz, 1999, <u>Post-Normal Science: Environmental Policy Under Conditions of Complexity</u>, http://www.jvds.nl/pns/pns.htm

Grey, S., 1995, <u>Practical Risk Assessment</u>, John Wiley & Sons, New York, NY

Grossman, T., 2001, "Causes of Decline", **INFORMS Transactions on Education**, 1, 2

Harvey, L.D., 2000, <u>Global Warming: The Hard Science</u>, Prentice Hall, New York, NY

Hammitt, J.K., and A.I. Shlaykhter, 1999, "The Expected Value of Information and the Probability of Surprise", **Risk Analysis**, 19, 1, p135-152

Henderson, S., 1989, "Species", **EPA Journal**, 15, p21-22

Hennessy, K.J., and A.B. Pittock, 1995, "Greenhouse Warming and Threshold Temperature Events in Victoria, Australia", **International Journal of Climatology**, 15, 591-612

Inch, J., 2001, "Characterizing Uncertainty: Methodology for Describing the Quality of Knowledge", Policy Research Directorate, Environment Canada, Draft: December 2001

IPCC, 1990, Climate Change: The IPCC Scientific Assessment, (Eds. Houghton, J.T., Jenkins, G.J. & Ephraums, J.J.). Cambridge University Press, Cambridge. 365pp.

IPCC, 1992, Climate Change 1992: The Supplementary Report to the IPCC Scientific Assessment, (Eds. Houghton, J.T., Callander, B.A. & Varney, S.K.). Cambridge University Press, Cambridge. 200pp.

IPCC, 2001, Climate Change 2001: The Scientific Basis. Contribution of Working Group I to the Third Assessment Report of the Intergovernmental Panel on Climate Change [Houghton, J.T., Ding, Y., Griggs, D.J., Noguer, M., van der Linden, P.J., Dai, X., Maskell, K. & Johnson, C.A. (eds.)]. Cambridge University Press, Cambridge, United Kingdom and New York, NY, USA, 881pp

Jones, C.V., 1991, Risk Analysis of Infrastructure: The Case of Water and Power, Quorum Books, Westport, CT

Katz, R.W., and B.G. Brown, 1992, "Extreme Events in a Changing Climate: Variability is More Important than Averages", **Climate Change**, 21, 289-302

Kempton, W., 1991, "Lay Perspectives on Global Climate Change", **Global Environmental Change: Human and Policy Dimensions**, 1, p183-208

Kettenburg, A., 1995, In: *Climate Change 1995: The Science of Climate Change, Contribution of Working Group 1 to the Second Assessment Report of the IPCC* (Eds. Houghton, J.T., Callander, B.A. & Varney, S.K.). Cambridge University Press, Cambridge, pp.285-357.

Kirkwood, C.W, and L.W. Seidman, 1985, "Avoiding Decision-Making Errors", **Pace**, 12, 9, p65-69

Kirkwood, C.W, 1997, Strategic Decision Making, Wadsworth Publishing Company, Belmont, CA

Kleijnen, J., and W. van Groenendaal, 1992, <u>Simulation: A Statistical Perspective</u>, John Wiley & Sons, New York, NY

Kleiner, A., 1994, "Creating Scenarios", in P.M. Senge, C. Roberts, R.B. Ross, B.J. Smith & A. Kleiner, <u>The Fifth Discipline Fieldbook: Strategies and Tools for Building a Learning Organization</u>, Doubleday Currency, New York, NY

Kleypas, J., and B. Opdyke, 1998, "Symposium Participants Assess Future of Coral Reefs", **EOS**, 79, p249, 251, 253

Law, A.M. and Kelton W.D., 1991, <u>Simulation Modeling and Analysis</u> (2nd edn). McGraw-Hill, New York, NY

Lee, R., 2001, "Climate Change and Environmental Assessment, Part 1: Review of Climate Change Considerations in Selected Past Environmental Assessments", Report to the Canadian Institute for Climate Studies, available on the CEAA web site: <u>http://www.ceaa-acee.gc.ca/0010/0001/0002/index_e.htm</u>

Leggett, J., W.J. Pepper, and R.J. Swart, 1992, "Emissions Scenarios for the IPCC: An Update", In: *Climate Change 1992: The Supplementary Report to the IPCC Scientific Assessment* (Eds. Houghton, J.T., Callander, B.A. & Varney, S.K.). Cambridge University Press, Cambridge, pp.69-95.

Loehle, C., 1996, "Do Simulations Predict Unrealistic Dieback", **Journal of Forestry**, 94, p13-15

Mackie, R, 2002, "Scientists Group Says Kyoto Rationale Flawed", *Globe & Mail*, November 20

Miller, G., 2002, <u>Climate Change: Is the Science Sound?</u>, Special Report to the Ontario Legislative Assembly, Environmental Commissioner of Ontario, Toronto, Ontario

Morgan, M.G. and D.W. Keith, 1995, "Subjective Judgments by Climate Experts", **Environmental Science and Technology**, 29, 468A-476A

Morgan, M.G., L. Pitelka, and E. Shevliakova, 2001, "Elicitation of Expert Judgements of Climate Change Impacts on Forest Ecosystems", **Climatic Change**, 49, p279-307

Moss, R.H. and S.H. Schneider, 2000, "Uncertainties in the IPCC TAR: Recommendations to Lead Authors for More Consistent Assessment and Reporting", Third Assessment Report Cross Cutting Issues Guidance Papers, IPCC, Switzerland, p33-51

Murphy, M., 2003, <u>A Review of the Consideration of Climate Change in Recent Environmental Assessments and Recommendations for Guidelines</u>, Master's Research Paper, Department of Geography, University of Toronto, Toronto, Ontario

Murtha, J.A., 1995, <u>Decisions Involving Uncertainty</u>, Palisade, Ithaca, NY

Nelson, F.E., O.A. Anisimov, and N.I. Shiklomanov, 2001, "Subsidence Risk From Thawing Permafrost", **Nature**, 410, p889-890

Normill, D., 2000, "Some Coral Bouncing Back from El Nino", **Science**, 288, p941-942

Palisade, 2002, <u>@Risk: Advanced Risk Analysis for Spreadsheets</u>, Palisade Corporation, Newfield, NY

Pounds, J.A., 2001, "Climate and Amphibian Declines", **Nature**, 410, p639-640

Ragsdale, C., 2001, <u>Spreadsheet Modeling & Decision Analysis</u>, 3rd ed., Thomson South-Western College Publishing, Cincinnati, OH

Ravetz, J.R., 1986, "Usable Knowledge, Usable Ignorance: Incomplete Science with Policy Implications", in Clark and Munn (eds.), <u>Sustainable Development of the Biosphere</u>, Cambridge University Press, New York, p415-432

Richards, D. and W.D. Rowe, 1999, "Decision-Making with Heterogeneous Sources of Information", **Risk Analysis**, 19, 1, p69-81

Roberts, L., 1988, "Is There Life After Climate Change", **Science**, 242, p1010-1012

Russo, J.E. and P.J.H. Schoemaker, 1989, <u>Decision Traps: Ten Barriers to Brilliant Decision-Making and How to Overcome Them</u>, Simon & Schuster, New York, NY

Schank, R.C., 1990, <u>Tell Me a Story: A New Look at Real and Artificial Memory</u>, Scribner, New York, NY

Schoemaker, P.J.H., 1995, "Scenario Planning: A Tool for Strategic Thinking", **Sloan Management Review**, 36, 2, p25-40

Schwartz, P., 1991, <u>The Art of the Long View</u>, Doubleday Currency, New York, NY

Showstack, R., 2000, "New Protection Initiatives for Coral Reefs", **EOS**, 82, p626, 628

Smit, B., and C. Yunlong, 1996, "Climate Change and Agriculture Change in China", in <u>Global Environment Change 6</u>, Elsevier Science Ltd., p205-214

Southam, C.F., B.N. Mills, R.J. Moulton, and D.W. Brown, 1999, "The Potential Impact of a Climate Change in Ontario's Grand River Basin: Water Supply and Demand Issues", **Canadian Water Resources Journal**, 24, 4, p307-330

SRES, 2000, "Special Report on Emission Scenarios", in <u>Summary for Policymakers: A Special Report of Working Group III</u>, ed: N. Nakicenovic *et al.*, Intergovernmental Panel on Climate Change, p1-20

Steffen, W., 2001, "Implications of Global Change for Natural and Managed Ecosystems: A Synthesis of GTCE and Related Research", **Global Change Newsletter**, 44, p1-2

Stirling, A., 1998, "Risk at a Turning Point", **Journal of Risk Research**, 1, 2, p102

Stone, R., 1995, "If the Mercury Soars, So May Health Hazards", **Science**, 267, p957-958

Taubes, G., 1995, "Is a Warmer Climate Wilting the Forests of the North?", **Science**, 267, March 17, p1595

Toronto Star, 2002a, "Scientists' Group Says Kyoto Rationale Flawed", November 13, 2002

Toronto Star, 2002b, "Kyoto a 'Goofily Concocted Theory': Klein", November 15, 2002

Toronto Star, 2002c, "Business Group 'Trying to Scare Canadians' on Kyoto", November 15, 2002

Toronto Star, 2002d, "Most energy-producing provinces blast Kyoto plan", November 21, 2002

van Asselt, M.B.A., R. Langendonck, F. van Asten, A. van der Giessen, P.H.M. Janssen, P.S.C. Heuberger, and I. Geuskens, 2001, "Uncertainty and RIVMs Environmental Outlooks: Documenting a Learning Process", RIVM, Bilthoven, p16-19

Wellington, M., 2000, "Crisis on Coral Reefs Linked to Climate Change", **EOS**, 82, p1, 5

Wigley, T.M.L., 1988, "The Effect of Changing Climate on the Frequency of Absolute Extreme Events", **Climate Monitoring**, 17, 44-55

Wilby, R.L. and Wigley, T.M.L., 1997, "Downscaling general circulation model output: a review of methods and limitations", **Progress in Physical Geography**, 21, 530-548.

Wilby, R.L., Hassan, H. and Hanaki, K., 1998a, "Statistical downscaling of hydrometeorological variables using general circulation model output", **Journal of Hydrology**, 205, 1-19

Wilby, R.L., Wigley, T.M.L., Conway, D., Jones, P.D., Hewitson, B.C., Main, J. and Wilks, D.S., 1998b, "Statistical downscaling of General Circulation Model output: a comparison of methods", **Water Resources Research**, 34, 2995-3008.

Wolfe, D.A., and J.D. Erickson, 1993, "Carbon Dioxide Effects on Plants: Uncertainties and Implications for Modeling Crop Response to Climate Change", in Agricultural Dimensions of Global Climate Change, eds. H.M. Kaiser and T.E. Drennen, St. Lucie Press, p153-178

WRI, 2000, <u>World Resources 2000-2001: People and Ecosystems, The Fraying Web of Life</u>, World Resources Institute, Washington DC

Wright, G. and P. Ayton, 1992, "Judgmental Forecasting in the Immediate and Medium Term", **Organizational Behavior and Human Decision Processes**, 51, 344-363

APPENDIX

Streamflow and Energy Model

The illustrative example of Chapter 4 is based upon a modified model of a run-of-the-river hydroelectric project that had been proposed for Northern Ontario. The EA for this proposed project contained considerable historical weather and streamflow data for the region and served to establish realistic values for the energy parameters related to the project design that were used in the example. However, since the focus of this study was on the incorporation and communication of climate change uncertainties into EAs and not on the design features of hydroelectric facilities, the underlying models that were assumed for both streamflow and energy production were quite basic.

The actual, proposed project contained 30 years of information on the monthly rates of historical streamflows. A hypothetical, no climate change, baseline condition extending over the period 2010-2099 was created consisting of three repetitions of this historical data. To apply climate change to the streamflows required the application of some form of climate modification model to these baseline conditions. While in reality any function used to model streamflow should not (or cannot) be treated as a simple function of weather data, the assumed future streamflow model employed a straightforward linear scaling/interpolation of the historical data based upon the future precipitation and temperature patterns within the various climate change scenarios. Namely, to generate future streamflows under the conditions of each scenario, the baseline streamflow pattern was modified proportionately to differences between historical temperature/precipitation data and the future projections of these variables as provided in the scenarios. Through this linear scaling process, historical streamflow rates were transformed into future streamflow patterns as projected under each of the different climate scenarios considered.

These projected future streamflow patterns then become inputs to the model used for energy generation at the hydro facility. In this energy model, the hypothetical run-of-the-river facility was assumed to operate without storage and at constant efficiency. To further simplify the analysis, energy production was assumed to vary at a linear rate with the streamflow, using the relationship expressed by:

Energy = Density *Average Head*Efficiency*Time*Streamflow.

The combined implications inherent in the above energy model simplifications are that considerable portions of the variability demonstrated in the Chapter 4 might, in fact, have been mitigated through the inclusion of appropriate storage at the facility. Hence, the inclusion of such a storage feature might reasonably be considered as one plausible alternative to consider during the hypothetical EA of the project.

For illustrative purposes of uncertainty and variability related to the incorporation of climate change into project EAs, these simplified streamflow and energy models were sufficient for their required instructive intentions. However, for any project actually addressed in an EA, significantly more attention to specific modelling and design details would be requisite.

0-595-33295-1